THE ILLUSTRATED FLORA OF ILLINOIS

The Illustrated Flora of Illinois

ROBERT H. MOHLENBROCK, General Editor

THE ILLUSTRATED FLORA OF ILLINOIS

FLOWERING PLANTS
magnolias to pitcher plants

Robert H. Mohlenbrock

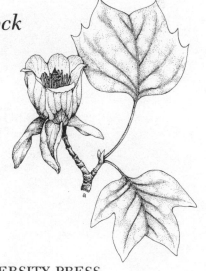

SOUTHERN ILLINOIS UNIVERSITY PRESS
Carbondale

Southern Illinois University Press
www.siupress.com

Cover illustration: *Aquilegia canadensis* (Columbine), by Miriam Meyer

Library of Congress Cataloging-in-Publication Data
Names: Mohlenbrock, Robert H., 1931– author.
Title: Flowering plants : magnolias to pitcher plants /
    Robert H. Mohlenbrock.
Other titles: Illustrated flora of Illinois.
Description: Paperback edition. | Carbondale : Southern Illinois
    University Press, 2017. | Series: The illustrated flora of Illinois |
    Includes index.
Identifiers: LCCN 2016044037 | ISBN 9780809335848 (pbk.) |
    ISBN 9780809380121 (e-book)
Subjects: LCSH: Dicotyledons—Illinois—Identification. |
    Dicotyledons—Illinois—Pictorial works. | Dicotyledons—Illinois—
    Geographical distribution—Maps.
Classification: LCC QK157 .M622 2017 | DDC 583.09773—dc23
LC record available at https://lccn.loc.gov/2016044037

This book is dedicated to
Marcella Elizabeth Kling,
my mother-in-law,
who has been interested in my work
for several years.

CONTENTS

ILLUSTRATIONS

FOREWORD

The dicotyledonous plants of Illinois will encompass several volumes of The Illustrated Flora of Illinois series. This is the third volume devoted to dicots. It follows publication of the ferns of Illinois, five volumes on the monocots of Illinois, and two previous volumes on dicots.

The Illustrated Flora of Illinois is a multivolumed flora of the state of Illinois, designed to present every group of plants, from algae and fungi through mosses, liverworts, and lichens, to ferns and seed plants. A description will be provided for each kind of plant known to occur in Illinois, along with illustrations showing the diagnostic features of each species. Distribution maps and ecological notes will be included. Keys to aid in easy identification of the plants will be presented.

An advisory board was created in 1964 to screen, criticize, and make suggestions for each volume of The Illustrated Flora of Illinois during its preparation. The board is composed of botanists eminent in their area of specialty—Dr. Gerald W. Prescott, University of Montana (algae); Dr. Constantine J. Alexopoulos, University of Texas (fungi); Dr. Aaron J. Sharp, University of Tennessee (bryophytes); Dr. Rolla M. Tryon, Jr., The Gray Herbarium (ferns); and Dr. Robert F. Thorne, Rancho Santa Ana Botanical Garden (flowering plants).

The author is editor of the series and will prepare many of the volumes. Specialists in various groups are preparing the volumes on plants of their special interest.

There is no definite sequence for publication of The Illustrated Flora of Illinois. Volumes will appear as they are completed.

The author expresses his deepest gratitude to the Joyce Foundation, whose generous support made possible the preparation of this volume.

Robert H. Mohlenbrock

Southern Illinois University
May 22, 1980

The Illustrated Flora of Illinois

magnolias to pitcher plants

County Map of Illinois

INTRODUCTION

Dicotyledons, or flowering plants which generally produce, upon germination, a pair of "seed leaves," called cotyledons, far outnumber the monocots, or single cotyledonous plants, in Illinois. This is the third volume of The Illustrated Flora of Illinois to be devoted to this great group of flowering plants, the dicots.

Botanists since the time of Linnaeus, and even before, have devised many schemes of classification for flowering plants. Some of these have been based on superficial, easy to recognize features, while others are based on more technical characters. As more is learned about the plants of the world through studies in cytology, biochemistry, embryology, and others, and with new tools available, such as the scanning electron microscope, views concerning relationships among plants change.

There is little wonder that the more recently proposed systems of classification, such as those of Thorne (1968), Cronquist (1968), and Hutchinson (1973), are completely different from Linnaeus's concepts in 1753 or, for that matter, from the widely known Engler system developed during the last part of the nineteenth century.

Because the interpretation of plant structures and their relative importance are viewed differently by each different botanist, even the contemporary systems have many dissimilarities.

In trying to decide what system of classification to follow in The Illustrated Flora, I have studied many different works. Although not agreeing entirely with his concepts, I have selected the system proposed in synoptical form in 1968 by Robert Thorne.

I departed from the Thorne system in the first dicot volume (Mohlenbrock, 1978) by separating the Hypericaceae from the Clusiaceae. In the present volume, I break from Thorne's classification by recognizing the Nymphaeaceae, Nelumbonaceae, and Cabombaceae as distinct families rather than subfamilies of the Nymphaeaceae.

Since the arrangement of orders and families proposed by Thorne is unfamiliar to many, an outline of the orders and families known to occur in Illinois is presented. Those names in boldface are described in this volume of The Illustrated Flora of Illinois.

Order Annonales
Family Magnoliaceae
Family Annonaceae
Family Aristolochiaceae
Family Calycanthaceae
Family Lauraceae
Family Saururaceae
Order Berberidales
Family Menispermaceae
Family Ranunculaceae
Family Berberidaceae
Family Papaveraceae
Order Nymphaeales
Family Nymphaeaceae
Family Nelumbonaceae
Family Cabombaceae
Family Ceratophyllaceae
Order Sarraceniales
Family Sarraceniaceae
Order Theales
Family Aquifoliaceae
Family Hypericaceae
Family Elatinaceae
Family Ericaceae
Order Ebenales
Family Ebenaceae
Family Styracaceae
Family Sapotaceae
Order Primulales
Family Primulaceae
Order Cistales
Family Violaceae
Family Cistaceae
Family Passifloraceae
Family Cucurbitaceae
Family Loasaceae
Order Salicales
Family Salicaceae
Order Tamaricales
Family Tamaricaceae
Order Capparidales
Family Capparidaceae
Family Resedaceae
Family Brassicaceae

Order Malvales
Family Sterculiaceae
Family Tiliaceae
Family Malvaceae
Order Urticales
Family Ulmaceae
Family Moraceae
Family Urticaceae
Order Rhamnales
Family Rhamnaceae
Family Elaeagnaceae
Order Euphorbiales
Family Euphorbiaceae
Order Solanales
Family Solanaceae
Family Convolvulaceae
Family Polemoniaceae
Order Campanulales
Family Campanulaceae
Order Santalales
Family Celastraceae
Family Santalaceae
Family Loranthaceae
Order Oleales
Family Oleaceae
Order Geraniales
Family Linaceae
Family Zygophyllaceae
Family Oxalidaceae
Family Geraniaceae
Family Balsaminaceae
Family Limnanthaceae
Family Polygalaceae
Order Rutales
Family Rutaceae
Family Simaroubaceae
Family Anacardiaceae
Family Sapindaceae
Family Aceraceae
Family Hippocastanaceae
Family Juglandaceae
Order Myricales
Family Myricaceae

Order Chenopodiales
Family Phytolaccaceae
Family Nyctaginaceae
Family Aizoaceae
Family Cactaceae
Family Portulacaceae
Family Chenopodiaceae
Family Amaranthaceae
Family Caryophyllaceae
Family Polygonaceae
Order Hamamelidales
Family Hamamelidaceae
Family Platanaceae
Order Fagales
Family Fagaceae
Family Betulaceae
Order Rosales
Family Rosaceae
Family Fabaceae
Family Crassulaceae
Family Saxifragaceae
Family Droseraceae
Family Staphyleaceae
Order Myrtales
Family Lythraceae
Family Melastomaceae
Family Onagraceae
Order Gentianales
Family Loganiaceae
Family Rubiaceae
Family Apocynaceae
Family Asclepiadaceae

Family Gentianaceae
Family Menyanthaceae
Order Bignoniales
Family Bignoniaceae
Family Martyniaceae
Family Scrophulariaceae
Family Plantaginaceae
Family Orobanchaceae
Family Lentibulariaceae
Family Acanthaceae
Order Cornales
Family Vitaceae
Family Nyssaceae
Family Cornaceae
Family Haloragidaceae
Family Hippuridaceae
Family Araliaceae
Family Apiaceae
Order Dipsacales
Family Caprifoliaceae
Family Adoxaceae
Family Valerianaceae
Family Dipsacaceae
Order Lamiales
Family Hydrophyllaceae
Family Boraginaceae
Family Verbenaceae
Family Phrymaceae
Family Callitrichaceae
Family Lamiaceae
Order Asterales
Family Asteraceae

This volume of The Illustrated Flora embraces four orders and fifteen families of plants. Since this is only a very small number of all the dicots in Illinois, no general key to the dicot families is presented in this book. Instead, the reader is invited to use my companion book, *Guide to the Vascular Flora of Illinois* (1975), for keys to all families.

The orders included in this work are the Annonales, Berberidales, Nymphaeales, and Sarraceniales. The fifteen families which comprise them are generally conceded by most botanists to be among the most primitive living plants in the world today.

The Annonales, containing the Magnoliaceae, Annonaceae, Aristolochiaceae, Calycanthaceae, Lauraceae, and Saururaceae, are considered first. Cronquist's system (1968) is similar, but he uses the order name Magnoliales for them and segregates the Aristolochiaceae into the Aristolochiales and the Saururaceae into the Piperales.

Both Thorne's classification and that of Cronquist place the Menispermaceae, Ranunculaceae, and Berberidaceae in the same order, Thorne calling it the Berberidales and Cronquist choosing to use the name Ranunculales. Thorne adds the Papaveraceae to this order, while Cronquist assigns the Papaveraceae to the Papaverales.

Thorne and Cronquist are in agreement as to the plants to be included in the Nymphaeales.

The four orders treated in this book can be characterized generally as woody in the Annonales (except for the Saururaceae and some Aristolochiaceae), herbaceous in the Berberidales (except for the Menispermaceae and some Berberidaceae), aquatic in the Nymphaeales, and insectivorous in the Sarraceniales.

The nomenclature for the species and lesser taxa used in this volume has been arrived at after lengthy study of recent floras and monographs. Synonyms, with complete author citation, which have applied to species in the northeastern United States, are given under each species. A description, while not necessarily intended to be complete, covers the more important features of the species.

The common name, or names, is the one used locally in Illinois. The habitat designation is not always the habitat throughout the range of the species, but only for it in Illinois. The overall range for each species is given from the northeastern to the northwestern extremities, south to the southwestern limit, then eastward to the southeastern limit. The range has been compiled from various sources, including examination of herbarium material and some field studies. A general statement is given concerning the range of each species in Illinois. Dot maps showing county distribution for each taxon are provided. Each dot represents a voucher specimen deposited in some herbarium. There has been no attempt to locate each dot with reference to the actual locality within each county.

The distribution has been compiled from field study as well as herbarium study. Herbaria from which specimens have been studied are the Field Museum of Natural History, Eastern Illinois University, the Gray Herbarium of Harvard University, Illinois Natural History Survey, Illinois State Museum, Knox College, Missouri Botanical Garden, the Morton Arboretum, New York Botanical Garden, Southern Illinois University, the United States National Herbarium, the University of Illinois, and Western Illinois University. In addition, a few private collections have been examined. The author expresses his gratitude to the curators and staffs of these herbaria who helped him in his study.

Each species is illustrated, depicting the habit and distinguishing features. The illustrations were prepared by Mrs. Miriam Wysong Meyer.

Miriam Meyer prepared all of the illustrations except 32, 57b, and 58a. These were done by my son, Mark Mohlenbrock. I appreciate their efforts. Mr. Douglas M. Ladd has assisted me in checking herbarium specimens and compiling distributional data. For his assistance, I am deeply grateful. Many thanks are also given to my wife, Beverly, who assisted me in several herbaria and who typed all the drafts of the manuscript.

DESCRIPTIONS AND ILLUSTRATIONS

Order Annonales

Thorne (1968), whose system of classification is being followed in The Illustrated Flora, considers the Annonales to be the most primitive of all extant flowering plants. Of the twenty-four families assigned to this order by Thorne, only six occur in Illinois, and one of them is introduced. Most of the other families are tropical, which is to be expected if it is believed that the very primitive angiosperms were tropical. The families of the order Annonales, in the sequence that they will be considered, are the Magnoliaceae, Annonaceae, Aristolochiaceae, Calycanthaceae, Lauraceae, and Saururaceae.

MAGNOLIACEAE–MAGNOLIA FAMILY

Trees or occasionally shrubs with alternate, simple leaves; stipules present; flowers usually solitary, perfect, actinomorphic; sepals and petals usually undifferentiated, free, often indefinite in number, usually spirally arranged; stamens numerous, free, spirally arranged; pistils numerous, usually free and crowded together on a receptacle; fruit usually a "cone."

This family, considered to be one of the most primitive of flowering plants, is composed of about a dozen genera.

KEY TO THE GENERA OF Magnoliaceae IN ILLINOIS

1. Leaves entire; petals without an orange blotch at the base within; seeds unwinged _____ 1. *Magnolia*
1. Leaves 4-lobed; petals with an orange blotch at the base within; seeds winged _____ 2. *Liriodendron*

1. *Magnolia* L.–*Magnolia*

Trees or occasionally shrubs with alternate, simple leaves, the leaf buds enclosed by the stipules; inflorescence of large, solitary flowers; flowers perfect; calyx and corolla usually similar in color, spirally arranged, composed of 3 free sepals and 6–9 free petals; sta-

6

mens numerous, usually with enlarged anthers and very short filaments; pistils many, usually borne spirally on an elongated receptacle; fruit a fleshy or dry "cone," with 1–2 berrylike seeds per carpel, hanging from slender threads.

Magnolias are thought to represent some of the most primitive flowering plants in the world by virtue of their numerous, free, spirally arranged parts and their enlarged anthers which open inward. Most members of this genus have large, showy flowers and make splendid ornamentals. Many species, varieties, and hybrids are in cultivation in Illinois. Particularly common in cultivation in southern Illinois are the saucer magnolia (*Magnolia soulangeana*) and the southern magnolia (*Magnolia grandiflora*).

Only a single native species occurs in Illinois.

1. **Magnolia acuminata** L. Syst. Nat. ed. 10, 2:1082. 1759. *Fig. 1.*

Tree to 30 m tall, with a trunk diameter up to 1 m and with a pyramidal crown; twigs stout, lustrous, reddish-brown; buds densely silky-hairy, 1.0–1.5 cm long; leaves alternate, simple, entire or slightly wavy along the margin, somewhat pubescent on the lower surface, oblong, broadest at or below the middle, short-acuminate at apex, cuneate at base, up to 25 cm long; flowers 4.0–6.5 cm across; sepals 3, greenish, smaller than the petals; petals 6–9, greenish-yellow, to 7 cm long, oblanceolate; stamens numerous; pistils numerous, free; fruiting "cone" 4–7 cm long, greenish at first, consisting of a cluster of coriaceous, dark red follicles; seeds 1–2 per follicle, bright red, suspended on a thread.

COMMON NAME: Cucumber Tree; Cucumber Magnolia.

HABITAT: Rich hardwood forests.

RANGE: New York to Ontario, south to Louisiana and Georgia.

ILLINOIS DISTRIBUTION: Confined to the southern two tiers of counties in Illinois and Jackson County.

The leaves of the cucumber tree somewhat resemble those of the tupelo (*Nyssa aquatica*) and the sour gum (*Nyssa sylvatica*), but differ by the presence of stipules. They also differ from the tupelo by their shorter petioles, and from the sour gum by their greater size of the blades.

The greenish-yellow flowers appear in early May and last only for a very short time. The fruits, which resemble small cucumbers

1. *Magnolia acuminata* (Cucumber Magnolia). *a*. Flowering branch, X½.

when immature, ripen their seeds in September.

In Illinois, cucumber tree is limited to rich hardwood forests where it is commonly associated with beech, tulip tree, and sugar maple.

Vasey (1870) was apparently the first to attribute this species to Illinois.

2. Liriodendron L.–Tulip Tree

Tall trees with alternate, simple, lobed leaves, the leaf buds covered by leathery scales and subtended by large stipules; inflorescence of large, solitary flowers; flowers perfect; sepals 3, reflexed; petals 6, free, in two rows; stamens numerous; pistils numerous, free, borne on an elongated receptacle; fruit a dry "cone," bearing samaralike seeds.

There are two species in this genus, one in China and the other in eastern North America.

1. Liriodendron tulipifera L. Sp. Pl. 535. 1753. Fig. 2.

Tree to 40 m tall, with a trunk diameter occasionally over 1 m; bark gray, smooth on young trees, becoming more rough, furrowed, and platy; twigs moderately stout, glabrous, yellowish to reddish-brown; buds to 1.5 cm long, with coriaceous bud scales; leaves alternate, simple, 2-lobed near the base, 2-lobed near the apex, occasionally with a large lobelike tooth on each of the 2 basal lobes, glabrous, broadly retuse at the apex, more or less truncate at the base, to 20 cm long, to 20 cm broad; petioles glabrous, to 8 cm long; flowers 4.0–5.5 cm across; sepals 3, greenish, reflexed; petals 6, pale yellow-green with orange bases, to 5 cm long; stamens numerous; pistils numerous, free, flat, narrow; fruiting "cone" to 7 cm long, falling in its entirety, composed of many winged seeds.

COMMON NAME: Tulip Tree; Tulip Poplar; Yellow Poplar.

HABITAT: Rich hardwood forests.

RANGE: Massachusetts to Ontario, south to Texas and Florida.

ILLINOIS DISTRIBUTION: Confined to the southern one-half of the state.

Tulip tree is one of the most majestic trees of the rich hardwood forests of southern Illinois. The tree is valuable in the timber industry because of its long, straight bole.

The distinctively shaped leaves sometimes possess an extra large tooth on the basal lobes.

The large flowers are generally inconspicuous because they are hidden among the dense foliage in May. The seeds ripen in September and often persist on the trees well into the winter.

The wood, which is soft, light, and easily worked, is much used for veneer, boxes, and lumber.

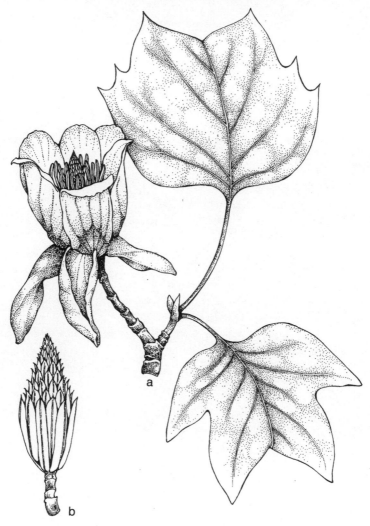

2. *Liriodendron tulipifera* (Tulip Tree).   *a*. Flowering branch, X⅜.   *b*. Fruiting "cone," X½.

ANNONACEAE–CUSTARD-APPLE FAMILY

Trees or shrubs with alternate, simple leaves; buds naked; stipules absent; flowers usually perfect, actinomorphic; sepals 3, free or united below; petals 6, usually free; stamens numerous, free; pistils (1–) numerous, free; fruits usually fleshy, with large seeds.

This predominantly tropical family is composed of about seventy-five genera.

Only the following genus occurs in Illinois.

1. *Asimina Adans.*–Pawpaw

Small trees with alternate, simple leaves; buds naked, elongated; flowers axillary, solitary; sepals 3, free; petals 6, free; stamens numerous, free; pistils 3–15; ovules numerous; fruit fleshy.

This strictly American genus is composed of about ten species. Kral (1960) has monographed the southeastern species.

Only the following species occurs in Illinois.

1. **Asimina triloba** (L.) Dunal, Monogr. Annon. 83. 1817. *Fig. 3.*

Annona triloba L. Sp. Pl. 537. 1753.
Porcelia triloba (L.) Pers. Syn. 2:95. 1807.
Uvaria triloba (L.) Torr. & Gray, Fl. N. Am. 1:45. 1838.

Small tree to 15 meters tall; young twigs pubescent, soon becoming more or less glabrous; buds narrow, dark rusty-brown, covered with golden hairs, up to 1.5 cm long, without scales; leaves obovate to ovate-oblong, acute to short-acuminate at the tip, cuneate or rounded at the base, up to 30 cm long, usually less than half as wide, glabrous; petioles up to 1 cm long, glabrous; flowers axillary, borne on shoots of the previous year, up to 3 (–4) cm across, appearing before the leaves have expanded fully; pedicels up to 3 cm long, pubescent; sepals 3, ovate, up to 1.2 cm long, early deciduous, pubescent; petals 6, in 2 series, maroon-purple, veiny, the outer ovate to orbicular, curving outward, up to 2.5 cm long, the inner ovate, erect, up to 2 cm long; stamens 6, free; pistils 3–several, but only one usually maturing; fruit an elongated berry, green at first, yellow when ripe, fleshy, pendulous, up to 15 (–18) cm long, up to 5 cm thick, with sweet, edible pulp; seeds flat, to 3 cm long.

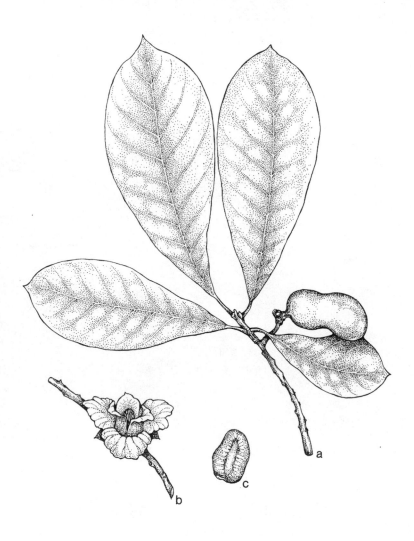

3. Asimina triloba (Pawpaw). *a.* Fruiting branch, X¼. *b.* Flower, X¾. *c.* Seed, X½.

COMMON NAME: Pawpaw.

HABITAT: Low woods and wooded slopes, often in alluvial soils.

RANGE: Ontario and western New York to Michigan, south to Nebraska, Texas, and Florida.

ILLINOIS DISTRIBUTION: Common in the southern counties, becoming less common northward.

The flowers of the pawpaw open in mid-April in southern Illinois and early May in the Chicago area. The flowers open as the leaves are beginning to expand. Flowering is finished by the end of May. The fruits, which ripen in September, are sweet and a delicacy for both wild animals and man.

Beck first reported this species from Illinois in 1828 as *Porcelia triloba* (L.) Pers. Mead in 1846 called it *Uvaria triloba* (L.) Torr. & Gray. Lapham (1857) was apparently the first in Illinois to adopt *Asimina triloba*.

Throughout all of Illinois, the pawpaw is commonly associated with sugar maple, beech, white ash, red oak, American elm, slippery elm, and basswood.

The wood of this species is light and soft and has little economic value.

ARISTOLOCHIACEAE–BIRTHWORT FAMILY

Herbs or shrubs, sometimes climbing; leaves basal or alternate; stipules absent; flowers perfect, actinomorphic or zygomorphic; calyx tubular, united at least at the base, 3-lobed; petals absent; stamens 6 or 12, closely appressed to the styles; ovary inferior or subinferior, 6-locular; fruit a capsule, many-seeded.

This family is composed of eight genera, all but *Asarum* and *Hexastylis* primarily tropical. About three-fourths of the 400 species in the family belong to the genus *Aristolochia*.

There are many botanists who place the Aristolochiaceae in its own order, the Aristolochiales.

KEY TO THE GENERA OF Aristolochiaceae IN ILLINOIS

1. Leaves 2, basal; stamens 12; calyx regular _____ 1. *Asarum*
1. Leaves more than 2, cauline; stamens 6; calyx irregular _____
   _____ 2. *Aristolochia*

1. Asarum L.—Wild Ginger

Perennial herbs from elongated rhizomes; leaves 2, basal, long-petiolate; flower solitary, borne on an elongated pedicel from the axil of the leaves; calyx united below into a short tube, actinomorphic, 3-lobed; petals absent; stamens 12, attached to the ovary; ovary inferior, 6-locular; fruit a capsule, dehiscing irregularly; seeds many.

Asarum is composed of about 60 species found in temperate and subtropical regions of the northern hemisphere.

Only the following variable species occurs in Illinois.

1. Asarum canadense L. Sp. Pl. 442. 1753.

Perennial herb from slender, branched, generally pubescent rhizomes; leaves 2, basal, reniform, acute to subacute at the apex, deeply cordate at the base, pubescent on both surfaces, entire, up to 12 cm long and broad at anthesis, expanding to 18 cm at maturity; petioles up to 15 cm long, densely pubescent; flower solitary, up to 2.5 cm across, pendulous from axil of the leaves, on a pubescent pedicel up to 5 cm long; calyx tube short, up to 10 mm long, adnate to the ovary, purple-brown, the 3 lobes deltoid to long-acuminate, recurved or strongly reflexed; stamens 12; ovary inferior; capsule coriaceous, up to 1.5 cm in diameter, crowned by the persistent calyx and stamens; seeds several, flattened.

Two varieties may be distinguished rather readily in Illinois.

1. Tips of the calyx lobes long-acuminate, 5–20 mm long, the lobes longer than the tube _____ 1a. *A. canadense* var. *canadense*
1. Tips of the calyx lobes short-pointed, up to 5 mm long, the lobes as long as the tube _____ 1b. *A. canadense* var. *reflexum*

1a. Asarum canadense L. var. canadense *Fig. 4a, b.*

Asarum canadense var. *acuminatum* Ashe, Contr. 1:2. 1897.
Asarum acuminatum (Ashe) Bickn. in Britt. & Brown, Ill. Fl. 3:513. 1898.

Tips of the calyx lobes long-acuminate, 5–20 mm long, the lobes longer than the tube.

4. Asarum canadense (Wild Ginger). var. *canadense*. *a*. Habit, X.45. *b*. Flower, X1.8. var. *reflexum*. *c*. Flower, X1.8.

COMMON NAME: Wild Ginger.

HABITAT: Rich woodlands, often in floodplains.

RANGE: Quebec to Minnesota, south to Missouri, Tennessee, and Virginia.

ILLINOIS DISTRIBUTION: Restricted to the northwestern one-half of the state.

Fernald (1950) and others recognize two varieties of *Asarum canadense* with long-acuminate calyx lobes, attributing both to Illinois. One of these, typical var. *canadense*, is said to have calyx lobes 1.0–2.5 cm long and narrowed to a slender tip 0.5–1.5 cm long. Variety *acuminatum* is recorded as having calyx lobes 1.5–3.5 cm long and tapered gradually to a slender tip. After examining considerable material of both varieties, I conclude there are not enough clear-cut differences to justify maintaining each of them.

In variety *canadense*, the calyx lobes are usually spreading but rarely reflexed.

This variety flowers in April and May.

1b. Asarum canadense L. var. **reflexum** l(Bickn.) B. J. Robins. Rhodora 10:32. 1908. *Fig. 4c.*

Asarum reflexum Bickn. Bull. Torrey Club 24:533. 1897.

Asarum reflexum var. *ambiguum* Bickn. Bull. Torrey Club 24:533. 1897.

Asarum canadense var. *ambiguum* (Bickn.) Farw. Rep. Mich. Acad. Sci. 20:173. 1918.

Tips of the calyx lobes short-pointed, up to 5 mm long, the lobes about as long as the tube.

COMMON NAME: Wild Ginger.

HABITAT: Rich woodlands.

RANGE: Connecticut to Michigan, south to Kansas, Missouri, Kentucky, and North Carolina.

ILLINOIS DISTRIBUTION: Common throughout Illinois.

I am combining two varieties (var. *reflexum* and var. *ambiguum*) under var. *reflexum*, but with some reluctance. Variety *reflexum*, as described by Bicknell, is clearly distinct from var. *canadense* by its short, reflexed tips of the calyx lobes.

An enigmatic variety is var. *ambiguum*, which seems to be intermediate between var. *canadense* and var. *re-*

flexum. It has calyx lobes 12–20 mm long and scarcely reflexed. Variety *reflexum* has calyx lobes only up to 12 mm long.

Swink (1974) indicates all specimens of *Asarum* in the Chicago area are referable to var. *reflexum*.

This plant is one of the most widely distributed wild flowers in Illinois. It begins to flower during April and may continue in bloom well into June.

2. *Aristolochia* L.–*Birthwort*

Perennial herbs or high climbers; leaves alternate, entire; flowers solitary (in the Illinois species), zygomorphic, perfect; calyx tubular, the tube curved, 3-lobed, adnate to the ovary, at least at the base; stamens 6, the anthers adnate to the stigmas; ovary partly inferior, 6-locular; fruit a capsule; seeds many.

Aristolochia is a genus of about 300 species. Two very diverse species occur in Illinois.

KEY TO THE SPECIES OF Aristolochia IN ILLINOIS

1. Erect herb; lower surface of leaves pubescent, but not white-woolly; calyx dark purple, to 15 mm long _____ 1. *A. serpentaria*
1. High climbing woody twiner; lower surface of leaves white-woolly; calyx yellow, with a purplish orifice, over 15 mm long ___ 2. *A. tomentosa*

1. Aristolochia serpentaria L. Sp. Pl. 961. 1753.

Perennial herb from a short, aromatic rhizome; stems erect, slender, to 35 (–50) cm tall, usually slightly pubescent; leaves extremely variable, ranging from linear-lanceolate to ovate, acuminate at the apex, cordate, sagittate, or hastate at the base, entire, slightly pubescent to glabrous on both surfaces, to 12 (–14) cm long, to 6 (–7) cm broad, the lowermost leaves reduced to scales; petioles to 2 cm long, glabrous or slightly pubescent; flower solitary from near the base of the stem, on slender pedicels, the pedicels up to 8 cm long, bearing small scales; calyx tubular, strongly curved, up to 1.5 cm long, pubescent, expanded at the apex into 3 blunt, purple-brown, shallow lobes; capsule subglobose to somewhat longer than broad, ridged, up to 1 cm in diameter; seeds obovoid, up to 5 mm long, with minute yellow-white warts.

Two varieties, rather distinct vegetatively, occur in Illinois.

1. Some or all the leaves over 2 cm across midway from base to apex _____ 1a. *A. serpentaria* var. *serpentaria*

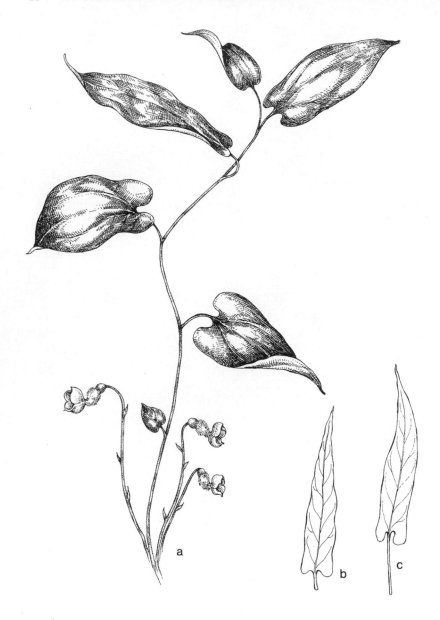

5. *Aristolochia serpentaria* (Virginia Snakeroot). var. *serpentaria*. *a*. Habit, with flowers, X1. var. *hastata*. *b*, *c*. Leaves, X1.

1. Leaves never more than 2 cm across midway from base to apex _ _ _ _ _
_ 1b. *A. serpentaria* var. *hastata*

1a. Aristolochia serpentaria L. var. **serpentaria** *Fig. 5a*.

Some or all the leaves over 2 cm across midway from base to apex.

COMMON NAME: Virginia Snakeroot.
HABITAT: Rich woodlands.
RANGE: Connecticut across Ohio to Kansas (*fide* Fernald, 1950), south to Texas and Florida.
ILLINOIS DISTRIBUTION: Local throughout the state, except for the northwestern counties where it is apparently absent.

The typical variety differs from var. *hastata* by its broader leaves with its mostly rounded basal lobes.

The flowers are borne so near the ground that they are often hidden by the leaf litter.

The roots, when bruised, have the strong aroma of turpentine.

After the capsule dehisces, it radiates into 6 lobes up to 4 cm across.

The flowers bloom from May to July, while the fruits mature in September.

1b. Aristolochia serpentaria L. var. **hastata** (Nutt.) Duchartre
in DC. Prod. 15(1):434. 1864. *Fig. 5b, c*.

Aristolochia hastata Nutt. Gen. N. Am. Pl. 2:200. 1818.

Aristolochia nashii Kearney, Bull. Torrey Club 21:485. 1894.

Leaves never more than 2 cm across midway from base to apex.

COMMON NAME: Narrow-leaved Snakeroot.
HABITAT: Rich or swampy woods.
RANGE: Virginia across to southeast Missouri, south to Texas and Florida.
ILLINOIS DISTRIBUTION: Known only from Alexander, Johnson, and Pulaski counties.

The leaves of this more southern variety are linear-lanceolate and sagittate to hastate at the base. On the basis of vegetative appearance, this taxon appears to be a distinct species, but there are no appreciable differences in the flower or fruit between this and the typical variety.

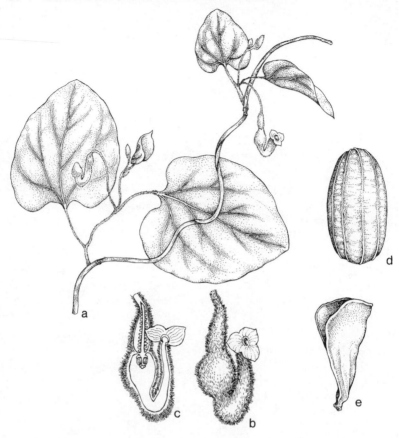

6. *Aristolochia tomentosa* (Dutchman's Pipevine). *a*. Habit, X¼. *b*. Flower, external, X¾. *c*. Flower, internal, X¾. *d*. Fruit, X⅜. *e*. Seed, X2.

The first collection of this variety from southern Illinois was made by Harry Ahles on October 5, 1942, from a cypress swamp five miles north of Grand Chain in Pulaski County. It also is known from several swamps in Johnson County, and from Alexander County.

2. Aristolochia tomentosa Sims, Bot. Mag. 33:pl. 1369. 1811.
Fig. 6.
Woody vine with twining, tomentose stems reportedly up to 30 m long; leaves alternate, broadly ovate to suborbicular, obtuse to sub-

acute at the apex, cordate at the base, entire, tomentose on both surfaces, up to 20 cm long, nearly as broad; petioles up to 6 cm long, tomentose; flowers solitary, axillary, on tomentose pedicels up to 4 cm long, without bracts; calyx tubular, strongly curved, tomentose, the tube yellow-green, up to 3 cm long, expanded at the tip into 3 acute, purple, spreading lobes; stigma 3-lobed, spreading; capsule cylindrical, tomentose, up to 8 cm long, up to 3 cm in diameter; seeds triangular, flat, notched at the apex.

COMMON NAME: Dutchman's Pipevine.

HABITAT: Chiefly calcareous woods.

RANGE: Indiana to Kansas, south to Texas and Florida.

ILLINOIS DISTRIBUTION: Confined to the southern half of Illinois.

The Dutchman's pipevine is a high-climbing woody vine with large, cordate, tomentose leaves. The common name alludes to the flower which has the calyx in the configuration of a Dutchman's pipe.

The woody fruits contain many flat seeds which are triangular in shape.

This species flowers in May and June.

A similar vine, A. durior Hill, reputedly cultivated in Illinois, differs mainly by its general lack of pubescence.

CALYCANTHACEAE–STRAWBERRY-SHRUB FAMILY

Shrubs with aromatic bark; leaves opposite, simple; stipules absent; flower solitary, perfect, actinomorphic; sepals and petals undifferentiated, indefinite in number, spirally arranged; stamens numerous, free, inserted on the receptacle; pistils several, free, usually enclosed by the hollow receptacle; fruit a fleshy receptacle containing several 1-seeded achenes.

The family has similarity to the Rosaceae because of the hollow receptacle and perigynous flowers. It also has been placed in the annonalian group by virtue of its numerous, spirally arranged parts.

Two genera, including the following, comprise the family. The other genus is native to China.

1. Calycanthus L.–Strawberry-shrub

Shrub with aromatic bark; leaves opposite, simple; flower solitary, purplish-red; sepals numerous, united below into a fleshy cup; pet-

7. *Calycanthus floridus* (Strawberry-shrub).   *a*. Flowering branch, X½.

als numerous, similar to the sepals; stamens numerous, the innermost sterile; pistils several, nearly enclosed in the hollow receptacle, the ovary superior, 1-locular; fruit a fleshy receptacle containing several 1-seeded achenes.

Four species make up this genus, all occurring in the United States.

Only the following escape from cultivation has been found in Illinois.

1. **Calycanthus floridus** L. Syst. ed. 10, 1066. 1759. *Fig. 7.*
Shrub to 3 m tall, the stems much branched, the twigs usually pubescent; leaves opposite, oval to ovate, obtuse to acute at the apex, cuneate to the base, entire, pubescent on both surfaces, to 5.5 cm long; flower solitary, axillary, purple, on a pubescent pedicel up to 3 cm long; sepals and petals similar, oblong to linear-oblong, obtuse to acute, pubescent, up to 2 cm long; stamens numerous; pistils several in the hollow receptacle; fruit obovoid, to 6 cm long, containing several seeds; seeds up to 1 cm long.

COMMON NAME: Strawberry-shrub.

HABITAT: Woodland.

RANGE: Virginia to West Virginia, south to Mississippi and Florida; escaped from cultivation in Illinois.

ILLINOIS DISTRIBUTION: Known from a single locality in Jackson County (Union Hill, four miles south of Carbondale).

This attractive species is an infrequently planted ornamental in southern Illinois. At its only station in Illinois, it has apparently spread from cultivation and persists on a wooded hillside.

The flowers, which smell like strawberries when crushed, open in May.

LAURACEAE–LAUREL FAMILY

Trees or shrubs, usually aromatic, with alternate, simple leaves; stipules absent; flowers bisexual or unisexual, actinomorphic; sepals 6 (in our species), free, in two whorls; petals absent; stamens 9 (in our species), the anthers with valvate dehiscence, the inner three each with usually a pair of glands at the base; pistil 1, the ovary superior, 1-locular; fruit a drupe, 1-seeded.

This primarily tropical family is composed of about 1,100 species. Other native genera in the United States, in addition to *Sassafras* and *Lindera*, are *Litsea, Persea, Ocotea, Nectandra, Licaria, Umbellularia*, and *Cassytha*. Genera of economic importance in this family are *Persea*, the avocado, *Cinnamomum*, the source of cinnamon and camphor, and *Ocotea*, the valuable greenheart wood. The true laurel of mythology is *Laurus nobilis*.

KEY TO THE GENERA OF Lauraceae IN ILLINOIS

1. Some of the leaves lobed; flowers appearing as the leaves unfold, in corymbose racemes; fruits blue _____ 1. *Sassafras*
1. None of the leaves lobed; flowers appearing before the leaves, in lateral clusters; fruits red _____ 2. *Lindera*

1. Sassafras Trew.–Sassafras

Aromatic trees or shrubs with alternate, simple, often lobed leaves; stipules absent; flowers unisexual, the plants dioecious; inflorescence corymbose-racemose, appearing with the leaves; sepals 6, free, persistent only in the pistillate flowers; petals absent; stamens 9, free, in three whorls, the inner three stamens each with a pair of stalked glands at the base of the filament; pistillate flowers with six small, rudimentary stamens; ovary superior, 1-locular; fruit a 1-seeded drupe, subtended by the persistent, cuplike calyx.

Our species and two in eastern Asia comprise this genus.
Only the following species occurs in Illinois.

1. Sassafras albidum (Nutt.) Nees, Syst. Laurin. 490. 1836.

Laurus sassafras L. Sp. Pl. 371. 1753.
Laurus albida Nutt. Gen. 1:259. 1818.
Sassafras sassafras (L.) Karst. Deuts. Fl. 505. 1881.

Small to large tree, sometimes attaining a height of 40 m and a diameter of nearly 5 m; bark rough, irregularly ridged; branchlets green, glabrous or pubescent; buds ovoid, acute to subacute, greenish, up to 6 mm long; leaves variable, often of three types on the same plant, unlobed, 2-lobed, or 3-lobed, entire, pubescent when young, sometimes becoming glabrous, up to 15 cm long, up to 10 cm across; petioles up to 2.5 cm long, glabrous or pubescent; flowers borne from the apex of the previous year's branches when the leaves first begin to unfold, greenish-yellow, up to 5 mm across; sepals 6, free, up to 2.5 mm long, linear to oblanceolate, usually incurved at the tip; petals absent; stamens about as long as the sepals; drupes blue, ellipsoid, up to 1.2 cm long, subtended by the red, persistent, cuplike calyx and red pedicels.

Variation in the amount of pubescence has led to the separation of two varieties.

KEY TO THE VARIETIES OF Sassafras albidum IN ILLINOIS

1. Leaves glabrous or nearly so on the lower surface at maturity _____
_____ 1a. *S. albidum* var. *albidum*
1. Leaves permanently pubescent on the lower surface _____
_____ 1b. *S. albidum* var. *molle*

1a. **Sassafras albidum** (Nutt.) Nees var. **albidum** *Fig. 8*.

Sassafras officinale Nees & Eberm. var. *albidum* (Nutt.) Blake, Rhodora 20:99. 1918.

Leaves glabrous or nearly so on the lower surface at maturity.

COMMON NAME: Sassafras; White Sassafras.
HABITAT: Dry or moist woods, thickets, roadsides.
RANGE: Maine to Michigan, south to Arkansas and Virginia.
ILLINOIS DISTRIBUTION: Common in the southern three-fourths of Illinois, rare or absent elsewhere.

The typical variety, sometimes referred to as white sassafras, has leaves glabrous or nearly so at maturity, even though the very young leaves and petioles are pubescent.

This variety and the following sprout readily and develop frequently in recently cutover areas, often forming thickets. They are particularly common along fencerows. Under optimum forest conditions, on the other hand, the sassafras may become a large tree. The largest sassafras in Illinois, and the second largest in the nation, occurs about three miles west of Carbondale, Jackson County. This mighty specimen is fifteen feet, two inches in circumference 4½ feet above the ground. In recent years, the top of this tree has been damaged by lightning. Swink (1974) indicates that sassafras is a pioneer species in disturbed sandy soils.

The leaves of the sassafras may be unlobed, or 2- or 3-lobed, mitten-shaped, all on the same tree. They give off a strong fragrance when crushed.

The pollen-producing staminate flowers are borne on trees separate from the pistillate flowers. In both cases the flowers appear

8. *Sassafras albidum* (Sassafras). *a*. Leaves, with fruits, X½. *b*. Staminate flower, X5. *c*. Pistillate flower, X5.

as the leaves just begin to unfold. The sepals remain attached in the pistillate flowers, while they fall off early in the staminate ones.

The bark of the roots may be steeped, and the resultant, refreshing tea in the past has had remedial powers attached to it.

Sassafras flowers during April and May.

1b. Sassafras albidum (Nutt.) Nees var. **molle** (Raf.) Fern. Rhodora 38:179. 1936.

Laurus variifolia Salisb. Prodr. 344. 1796, *nom. illegit*.
Sassafras trilobum Raf. var. *molle* Raf. Aut. Bot. 85. 1840.
Sassafras officinale Nees & Eberm. Handb. Med.-Pharm. Bot. 2:418. 1831.
Sassafras variifolium (Salisb.) Kuntze, Rev. Gen. Pl. 2:574. 1891.

Leaves permanently pubescent on the lower surface.

COMMON NAME: Red Sassafras.
HABITAT: Woodlands, thickets, roadsides.
RANGE: Maine to Kansas, south to Texas and Florida.
ILLINOIS DISTRIBUTION: Occasional in the southern three-fourths of Illinois.

Mature leaves of this variety are pubescent on the lower surface, and this pubescence persists until the leaves fall.

Other differences between var. *molle* and the typical variety appear to be nonexistent, although some native folks insist there is a difference in the quality of the roots for tea.

The vegetative pubescence is not totally clear-cut since young leaves of the typical variety are usually pubescent.

2. *Lindera* Thunb.–Spicebush

Aromatic trees or shrubs with alternate, simple leaves; stipules absent; flowers unisexual, the plants dioecious; inflorescence in sessile clusters, appearing before the leaves; sepals 6, free, deciduous; petals absent; stamens 9, free, in three whorls, the inner three stamens each glandular at the base of the filament; pistillate flowers with up to 18 small rudimentary stamens; ovary superior, 1-locular; fruit a 1-seeded drupe.

Nearly sixty species, predominantly found in eastern Asia, comprise this genus. All of them have aromatic properties.

Only the following species occurs in Illinois.

1. Lindera benzoin (L.) Blume, Mus. Bot. Ludg.-Bat. 1:324. 1851.

Laurus benzoin L. Sp. Pl. 371. 1753.
Benzoin benzoin (L.) Coulter, Mem. Torrey Club 5:164. 1894.

Shrub to 5 m tall, much branched; bark smooth, brown; branchlets dark brown, glabrous, or the new growth sometimes pubescent; leaves elliptic to oblong-obovate, acute to short-acuminate at the apex, cuneate at the base, entire, glabrous or pubescent beneath, pale on the lower surface, to 12 cm long, up to half as wide; petiole up to 1.2 cm long, glabrous or pubescent; flowers in nearly sessile clusters from nodes of the previous season; flowers yellow, aromatic, up to 3 mm across, appearing before the leaves; sepals 6, free, about 1.5 mm long, linear to linear-lanceolate; petals absent; stamens as long as or a little longer than the sepals; drupes red, ellipsoid, up to 1 cm long, up to 6 mm in diameter.

Two varieties may be recognized in Illinois.

1. Leaves and young branchlets glabrous _____
_____ 1a. *L. benzoin* var. *benzoin*
1. Leaves and young branchlets pubescent _____
_____ 1b. *L. benzoin* var. *pubescens*

1a. Lindera benzoin (L.) Blume var. **benzoin** *Fig.* 9

Laurus aestivalis L. Sp. Pl. 370. 1753.

Benzoin aestivale (L.) Nees, Syst. Laurin. 495. 1836.

Leaves and young branchlets glabrous.

COMMON NAME: Spicebush.

HABITAT: Rich woodlands.

RANGE: Maine and Ontario to Michigan, south to Kansas, Kentucky, and North Carolina.

ILLINOIS DISTRIBUTION: Common throughout the state, except the northwestern portion.

The typical variety of *L. benzoin* is common in most of Illinois, where it grows in rich woodlands. It is frequently associated with the bladdernut (*Staphylea trifolia*) and the usual early spring wild flowers of rich woods. Common overstory trees are sugar maple, beech, and white ash. This plant shows a range in Illinois similar to that of the sassafras, also a member of the Lauraceae.

All parts of the plant yield a strong, spicy odor when crushed. The leaves of young shoots and sprouts are considerably larger than those produced on the mature branches.

The yellow flowers are borne on leafless branchlets from mid-March in southern Illinois until mid-May in northern Illinois. The red berries ripen in August and September.

9. *Lindera benzoin* (Spicebush). *a*. Leafy branch, X½. *b*. Flowering branch, X½. *c*. Fruiting twig, X¼. *d*. Staminate flower, X12½.

1b. Lindera benzoin (L.) Blume var. **pubescens** (Palmer and Steyerm.) Rehder, Journ. Arn. Arb. 20:412. 1939.

Benzoin aestivale (L.) Nees var. *pubescens* Palmer & Steyerm. Ann. Mo. Bot. Gard. 22:545. 1939.

Leaves and young branchlets pubescent.

COMMON NAME: Hairy Spicebush.

HABITAT: Swampy woods.

RANGE: New Jersey to Michigan, south to Texas and Florida.

ILLINOIS DISTRIBUTION: Known only from Jackson, Johnson, and Union counties.

The hairy spicebush differs from typical var. *benzoin* by its pubescent leaf surfaces and young branchlets. There seem to be no other differences which can correlate with the pubescence. Variety *pubescens*, at least in Illinois, is confined to more swampy situations than var. *benzoin*.

Ridgway (1872), Tehon (1942), and others have attributed the very rare, southeastern *Lindera melissaefolium* (Walt.) Blume to southern Illinois. Although *L. melissaefolium* is known from southeastern Missouri, it has not apparently been found in Illinois. Earlier reports of it are based on *L. benzoin* var. *pubescens*. While the pubescence of the two is somewhat similar, *L. melissaefolium* differs by its rounded or cordate leaf bases and its larger fruits.

Lindera benzoin var. *pubescens* flowers in March and April.

SAURURACEAE–LIZARD'S-TAIL FAMILY

Perennial herbs, with alternate, simple, entire leaves; flowers perfect (in our genus), small, actinomorphic; perianth absent; stamens up to 8; pistils 3–4, free or united at the base, 1-locular; fruit separating into 3 or 4 1-seeded segments.

In addition to *Saururus*, which has three species in eastern Asia and ours, three other monotypic genera belong to this family. One of them, *Anemopsis*, occurs in California and the southwestern United States.

Only the following genus occurs in Illinois.

1. Saururus L.–Lizard's-tail

Perennial herbs; leaves alternate, simple, entire, cordate, the petioles sheathing at the base; stipules absent; flowers perfect, small, in spikes, bractlets present; perianth absent; stamens 6–8, free; pistils 3–4, united at the base; fruit separating in 3–4 1-seeded segments.

Because of its great reduction of flower parts, this genus traditionally has been classified as one of the most primitive dicotyledonous genera in the world. In floras which adhere to the Engler system of classification, *Saururus* is usually the first dicot genus treated. The position relegated to it in The Illustrated Flora indicates an advanced, reduced condition.

Only the following species occurs in Illinois. Three other species grow in Asia.

1. Saururus cernuus L. Sp. Pl. 341. 1753. *Fig. 10.*

Perennial herb with stolons and slender aromatic rhizomes; stems erect, jointed, pubescent at first, usually becoming glabrous or nearly so at maturity, up to about 1 m tall, often branched; leaves alternate, ovate, acute at the apex, cordate at the base, entire, pubescent when young, becoming glabrous, palmately 5- to 9-nerved, up to 15 cm long, up to 8 cm broad, the petioles densely sheathing at the base; spikes 1–2, terminal but often overtopped by the upper leaves, up to 15 cm long, on peduncles up to 8 cm long, the spike strongly hooked or nodding toward the apex; flowers white, sessile or short-pedicellate, subtended by a bractlet, fragrant; perianth absent; stamens 6–8, free, up to 4 mm long, longer than the pistils; pistils 3–4, united at the base, up to 2 mm long, each with a recurved stigma; fruits separating at maturity, each segment up to 3 mm in diameter, strongly rugose.

COMMON NAME: Lizard's-tail.

HABITAT: Swampy woods, sometimes in standing water; wet ditches.

RANGE: Quebec and New York to Minnesota, south to Texas and Florida.

ILLINOIS DISTRIBUTION: Occasional to common in the southern two-thirds of Illinois, much less common northward.

Lizard's-tail is a characteristic species of the swamps of southern Illinois; it frequently occurs in sizable colonies. It sometimes grows in shallow standing water, particularly in roadside ditches.

10. Saururus cernuus (Lizard's-tail).   *a*. Flowering branch, X.45.   *b*. Flower, X5.4.
c. Stamen, X11¼.   *d*. Pistils, X9.   *e*. Cluster of fruits, X4½.

The white color of the flowers is due to the stamens since there is no perianth. The flowers, densely crowded into spikes, have a mild, pleasant fragrance. The 3–4 pistils in each flower are barely united at the base, but as the fruits mature, they become separated. Each fruiting segment, which has only one seed, is strongly wrinkled.

Lizard's-tail flowers from May to September.

Order Berberidales

This order is considered in the Thorne system of classification to be advanced over the Annonales. While most families assigned to the Annonales are trees, the families in the Berberidales are composed mostly of shrubs and herbs. In Illinois, the Berberidales contains the Menispermaceae, Ranunculaceae, Berberidaceae, and Papaveraceae (including Fumariaceae).

MENISPERMACEAE–MOONSEED FAMILY

Woody or herbaceous vines without tendrils; leaves alternate, simple; stipules absent; flowers dioecious, small, arranged in panicles, racemes, or cymes; sepals 3–9 (–12), in 2 whorls, free; petals 0, 3, or 6 (–12), in two whorls, free, usually smaller than the sepals; stamens 3, 6, 9, 12, or many, usually free; pistils (1–2) 3, or 6, free, the ovary superior, 1-locular; fruit a drupe; seeds often curved.

Because of the presence of free pistils, a condition known as apocarpy, this family is thought by some to be rather primitive. On the other hand, the Menispermaceae has several features deemed to be advanced—viny habit, absence of petals (sometimes), and dioecism.

This family if vines is chiefly tropical, composed of about seventy genera. Only three genera occur in northeastern North America, and each is known from Illinois. Several of the tropical species in this family are poisonous and are used by various aborigines as arrow poisons.

KEY TO THE GENERA OF Menispermaceae IN ILLINOIS

1. Leaves deeply 3- to 7-lobed; petals none; drupe 15–25 mm long _____ 1. *Calycocarpum*
1. Leaves entire or with 3–7 angles or shallow lobes; petals 6–8; drupe 5–10 mm long _____ 2
   2. Drupe blue-black; stamens 12–24; seed not snail-shaped _____ 2. *Menispermum*
   2. Drupe red; stamens 6; seed snail-shaped _____ 3. *Cocculus*

1. *Calycocarpum Nutt.*–Cupseed

High-climbing woody vine; leaves alternate, palmately lobed; infloresence arising from above the leaf axils, racemose-paniculate, usually pendulous; staminate flowers with 6 free sepals, o petals, 12 free stamens; pistillate flowers with 6 free sepals, o petals, 3 free pistils; fruit a drupe.

Calycocarpum is one of the few genera in the family which does not have petals.

Only the following species comprises the genus.

1. **Calycocarpum lyonii** (Pursh) Gray, Gen. Fl. Am. Illus. 1:76. 1848. *Fig. 11*.

Menispermum lyoni Pursh, Fl. Am. Sept. 371. 1814.

Woody climber; stems glabrous or sparsely pubescent; leaves alternate, ovate to orbicular in outline, usually 3- to 7-lobed but entire, cordate at the base, each lobe acute to acuminate, glabrous above, pubescent below, at least on the veins, to 20 cm long, nearly as broad; petiole about as long as the blade, glabrous or sparsely pubescent; inflorescence racemose-paniculate, pendulous, up to 25 cm long; flowers numerous, greenish-white, up to 4 mm across; sepals 6, free, oblong, obtuse, up to 2 mm long; petals absent; stamens 12, free, up to 2 mm long; pistils 3, free, up to 2 mm long, with the stigma radiating into several lobes; drupe black, oval, up to 2.5 cm long, 1-seeded, the stone flattened, hollowed out on one side.

COMMON NAME: Cupseed.

HABITAT: Swampy woodlands.

RANGE: Southern Indiana to eastern Kansas, south to Louisiana and Florida.

ILLINOIS DISTRIBUTION: Confined to the southern one-fifth of the state.

Cupseed is a rather rare climber of swampy woodlands in the southern three tiers of counties in Illinois. There is considerable variation in the degree of lobing of the leaf. Sometimes the leaves resemble those of our native grapes.

The common name of cupseed is derived from the seed which is flat and deeply hollowed out on one side.

This species flowers in May and June. The black drupes ripen in late summer.

11. Calycocarpum lyonii (Cupseed). *a*. Flowering branch, X½. *b*. Staminate flower, X6. *c*. Fruit, X1.

2. *Menispermum* L.–*Moonseed*

Climbing woody vines; leaves alternate, shallowly lobed or entire, peltate; inflorescence arising from above the leaf axils, paniculate; staminate flowers with 4–8 free sepals, 4–8 free petals, 12–24 free stamens; pistillate flowers with 4–8 free sepals, 4–8 free petals, 2–4 free pistils; fruit a drupe.

In addition to the species enumerated below, there is a second species native to eastern Asia.

Only the following species occurs in Illinois.

1. Menispermum canadense L. Sp. Pl. 340. 1753. *Fig. 12*.

Woody climber; stems glabrous or sparsely pubescent; leaves alternate, ovate, entire or shallowly 3- to 7-lobed or -angled, obtuse to acute to acuminate at the apex, cordate or rounded at the base, glabrous above, usually glabrous and paler below, up to 15 cm long, nearly as broad, peltate; petiole slender, as long as or slightly longer than the blade, glabrous or sparsely pubescent; inflorescence paniculate, up to 15 cm long; flowers numerous, whitish, up to 4 mm across; sepals usually 6, free, lanceolate, acute, up to 2 mm long; petals usually 6, free, lanceolate, acute, up to 2 mm long; petals usually 6, free, lanceolate, acute, shorter than the sepals; stamens 12–24, free, about as long as the sepals; pistils 2–4, free, often surrounded by six staminodia; drupe bluish-black, globose, up to 8 (–10) mm in diameter, 1-seeded, the stone flattened and ridged.

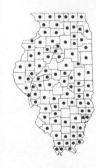

COMMON NAME: Moonseed.

HABITAT: Moist woodlands; thickets.

RANGE: Quebec to Manitoba, south to Oklahoma and Georgia.

ILLINOIS DISTRIBUTION: Common throughout the state.

Moonseed is one of the more common vines in the state, although it is relatively inconspicuous. It rarely climbs to the tops of trees as *Calycocarpum lyonii* does. This species, because of the leaf shape, resembles the small yellow passion flower (*Passiflora lutea* L.), but differs in the absence of tendrils and in its peltate leaves.

The clusters of fruits are reminiscent of grapes, but are not edible and may even be poisonous. The common name is derived from the moon-shaped seeds.

The flowers bloom from May to July.

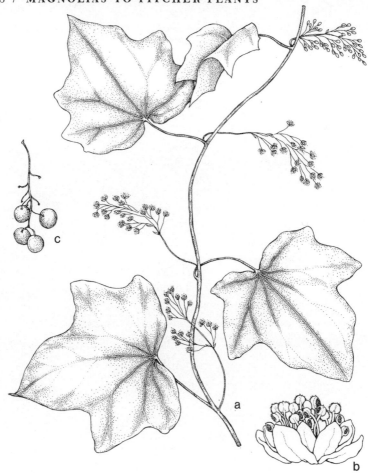

12. *Menispermum canadense* (Moonseed). *a*. Habit, X½. *b*. Staminate flower, X7½. *c*. Cluster of fruits, X½.

3. *Cocculus* DC.–Snailseed

Climbing woody or herbaceous vines; leaves alternate, entire or lobed; inflorescence arising from above the leaf axils, paniculate or racemose; staminate flowers with 6 free sepals, 6 free petals, 6 free stamens; pistillate flowers with 6 free sepals, 6 free petals, 3–6 free pistils; fruit a drupe.

There are about twelve species in this genus, most of which are native to the tropics.

Only the following species occurs in Illinois.

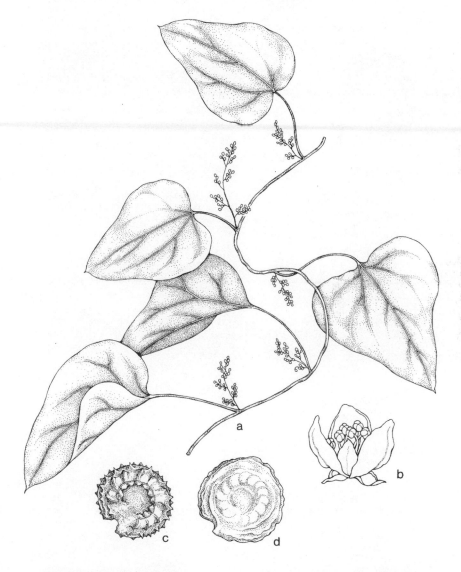

13. *Cocculus carolinus* (Snailseed).   *a*. Flowering branch, X½.   *b*. Staminate flower, X12½.   *c, d*. Seeds, X5.

1. **Cocculus carolinus** (L.) DC. Syst. Veg. 1:515. 1817. *Fig. 13.*
Menispermum carolinum L. Sp. Pl. 340. 1753.
Epibaterium carolinum (L.) Britt. in Britt. & Brown, Ill. Fl. 2:131. 1913.

Woody climber; stems glabrous or sometimes pubescent; leaves alternate, ovate to deltoid, entire or very shallowly and irregularly lobed, obtuse to acute at the apex, cordate to truncate at the base, glabrous above, glabrous or densely pubescent below, up to 12 cm long, often nearly as broad; petioles up to 10 cm long, glabrous or sometimes pubescent; inflorescence paniculate, up to 20 cm long; flowers numerous, greenish, up to 2 mm across; sepals 6, free, very broadly lanceolate, acute to subacute, up to 2 mm long; petals 6, free, concave, shorter than the sepals; stamens 6, free, usually a little shorter than the petals; pistils 3–6, free, often surrounded by staminodia; drupe red, flattened on both sides, up to 6 mm in diameter, 1-seeded, the stone curved into a spiral resembling a snail, irregularly ridged.

COMMON NAME: Snailseed.
HABITAT: Moist woods; thickets.
RANGE: Virginia to southeastern Kansas, south to Texas and Florida.
ILLINOIS DISTRIBUTION: Confined to the southern one-third of the state where it is occasionally found.

This species has clusters of beautiful red fruits during August to October. The stone within the fruit is coiled into a spiral resembling a snail.

There is considerable variability in leaf shape and pubescence. Several Illinois specimens are densly pubescent on the lower surface of the leaves. Some plants bear no lobed leaves at all, while others may have leaves irregularly shallow-lobed.

This is primarily a southern species which is fairly abundant in the southern tiers of counties. Its range extends northward along the Mississippi River to St. Clair County and along the Wabash River to Wabash County.

This species flowers during July and August.

RANUNCULACEAE–BUTTERCUP FAMILY

Chiefly perennial herbs (vine in *Clematis*); leaves primarily alternate (opposite in *Clematis*), simple or compound; stipules absent;

flowers usually perfect, actinomorphic or zygomorphic; sepals 3–15, free, often petallike; petals up to 15, or absent, usually free; stamens numerous, spirally arranged, free; pistils few to several, rarely 1, the ovary superior, 1-locular; fruit an achene, follicle, or berry.

The family contains about forty genera and approximately 1,500 species, mostly in the north temperate regions. Certain genera, such as *Hydrastis*, show a striking similarity to the Berberidaceae.

The eighteen genera which occur in Illinois fall into three tribes. Tribe Helleboreae is characterized by the carpels usually with two or more ovules, the fruit a follicle or berry, and the sepals imbricate in bud. Illinois genera in this tribe are *Caltha, Delphinium, Actaea, Cimicifuga, Hydrastis, Isopyrum, Aquilegia, Helleborus, Nigella,* and *Eranthis.* Tribe Anemoneae has carpels with one ovule, fruit an achene, and the sepals imbricate in bud. In this tribe are the genera *Ranunculus, Trautvetteria, Thalictrum, Hepatica, Anemonella, Anemone,* and *Myosurus.* Tribe Clematideae, comprised only of *Clematis,* has the sepals valvate in bud. It also differs by having opposite leaves.

Several members of the family are grown as garden ornamentals. Others are important because of their medicinal properties.

KEY TO THE GENERA OF Ranunculaceae IN ILLINOIS

1. Flowers yellow or white or occasionally pinkish _____ 2
1. Flowers red, blue, purple, or green _____ 16
  2. Stems climbing; leaves opposite _____ 18. *Clematis*
  2. Stems erect or creeping or floating; leaves basal or alternate ___ 3
3. Flowers yellow _____ 4
3. Flowers white or occasionally pinkish _____ 6
  4. Sepals green, petals yellow; fruit an achene _____ 1. *Ranunculus*
  4. Sepals yellow; petals reduced to 2-lipped nectaries or absent; fruit a follicle _____ 5
5. Petals absent; leaves simple, crenate _____ 2. *Caltha*
5. Petals reduced to 2-lipped nectaries; leaves palmately cleft _____ 17. *Eranthis*
  6. Plants aquatic; sepals and petals each 5, differentiated _____ 1. *Ranunculus*
  6. Plants not true aquatics; petals absent or, if present, either stamenlike or with a spur _____ 7
7. Flowers spurred _____ 3. *Delphinium*
7. Flowers without a spur _____ 8
  8. Flowers numerous in racemes, panicles, or corymbs; sepals incon-

spicuous, falling away as the flower opens _____ 9

8. Flowers 1–4, never in a raceme; sepals showy, petallike, persistent _____ 12

9. Leaves simple, although deeply lobed _____ 4. *Trautvetteria*

9. Leaves variously compound _____ 10

   10. Flowers unisexual, arranged in much-branched panicles _____ 5. *Thalictrum*

   10. Flowers perfect, arranged in racemes or corymbs _____ 11

11. Fruit fleshy, berrylike; raceme simple _____ 6. *Actaea*

11. Fruit dry, follicular; raceme sparingly branched ____ 7. *Cimicifuga*

   12. Leaves 3-lobed, all basal _____ 8. *Hepatica*

   12. Leaves either not 3-lobed or not all basal _____ 13

13. Leaves on the stem alternate _____ 14

13. Leaves on the stem opposite or whorled _____ 15

   14. Leaves ternately compound; sepals 5; pedicels glabrous _____ 10. *Isopyrum*

   14. Leaves simple, palmately lobed; sepals 3; pedicels pubescent- --- 9. *Hydrastis*

15. Leaves ternately compound; roots tuberous-thickened _____ 11. *Anemonella*

15. Leaves deeply to shallowly palmately lobed, but not ternately divided; plants rhizomatous, or with a woody caudex _____ 12. *Anemone*

   16. Leaves all basal _____ 17

   16. At least some of the leaves cauline _____ 18

17. Leaves linear, entire; sepals spurred; flowers greenish; receptacle elongated _____ 13. *Myosurus*

17. Leaves palmately lobed; sepals not spurred; flowers pinkish or light purplish; receptacle not elongated _____ 8. *Hepatica*

   18. One or more petals or sepals prolonged backward into a spur 19

   18. None of the perianth parts spurred _____ 20

19. Flowers red and yellow, with each of the five petals prolonged backward into a long spur _____ 14. *Aquilegia*

19. Flowers purple, blue, or green; one of the petallike sepals prolonged backward into a spur _____ 3. *Delphinium*

   20. Cauline leaves opposite or whorled _____ 21

   20. Cauline leaves alternate _____ 22

21. Sepals 4, thick and fleshy; leaves opposite _____ 18. *Clematis*

21. Sepals usually 5–20, thin; leaves whorled _____ 12. *Anemone*

   22. Flowers green _____ 15. *Helleborus*

   22. Flowers purplish or blue _____ 23

23. Flowers blue; inflorescence subtended by a deeply divided involucre;

flowers perfect _____ 16. *Nigella*
23. Flowers purplish; inflorescence without an involucre; flowers
unisexual _____ 5. *Thalictrum*

1. *Ranunculus* L.–*Buttercup*

Perennial herbs with fibrous or sometimes thickened roots; stems
erect or creeping and rooting at the nodes, sometimes hollow;
leaves usually both basal and cauline, simple and entire or toothed
or lobed, or compound; inflorescence axillary or terminal; flowers
perfect, actinomorphic; sepals usually 5, spreading or reflexed; petals (1–) 5 (–numerous in "double-flowered" forms), yellow or rarely
white; stamens (3–) 10– numerous; pistils (5–) 10–numerous; achenes crowded into fruiting heads, plump or flattened, smooth or
papillate, beaked.

There are about 250 species of *Ranunculus* distributed widely
in temperate or subarctic areas. In the tropics, this genus is generally restricted to the mountains. In the northeastern United States,
Fernald (1950) attributes thirty-six species. Twenty-five species are
recognized for Illinois, with nineteen considered to be native members of the flora.

In order to make positive identification of most species of *Ranunculus*, it is necessary to have mature achenes available. Reliance
on vegetative characters is not practical in distinguishing *R. abortivus* from *R. micranthus*, or in distinguishing members of the *R.
hispidus–R. fascicularis–R. septentrionalis–R. carolinianus* com-
plex.

A few taxa, such as *R. trichophyllus*, *R. longirostris*, *R. flabellaris*, and *R. gmelinii* var. *hookeri*, grow in water. Most other taxa
are associated with low, wet soil. At the other extreme are *R. rhomboideus*, a prairie species, and *R. micranthus* and *R. hispidus*, species of usually dry woodlands.

Benson's treatment in 1948 is the most recent extensive one on
this genus in North America. He attributes ninety-six species to
North America, divided among nine subgenera and numerous sec-
tions. The twenty-five species in Illinois are classified by Benson as
follows:

SUBGENUS *Ranunculus*
    Ranunculus(§ *Chrysanthe* [Spach] L. Benson)
    Ranunculus repens L.
    Ranunculus bulbosus L.
    Ranunculus acris L.

Ranunculus recurvatus Poir. in Lam.
Ranunculus pensylvanicus L. f.
Ranunculus septentrionalis Poir. in Lam.
Ranunculus carolinianus DC.
Ranunculus hispidus Michx.
Ranunculus fascicularis Muhl. ex Bigel.
§ *Echinella* DC.
Ranunculus parviflorus L.
Ranunculus sardous Crantz
Ranunculus arvensis L.
§ *Epirotes* (Prantl) L. Benson
Ranunculus rhomboideus Goldie
Ranunculus harveyi (Gray) Britt.
Ranunculus micranthus Nutt. ex Torr. & Gray
Ranunculus abortivus L.
§ *Flammula* (Webb) Ruoy & Foucand
Ranunculus ambigens S. Wats.
Ranunculus laxicaulis (Torr. & Gray) Darby
Ranunculus pusillus Poir. ex Lam.
§ *Hecatonia* (Lour.) DC.
Ranunculus sceleratus L.
Ranunculus gmelinii DC.
Ranunculus flabellaris Raf. apud Bigel.
SUBGENUS *Cyrtorhyncha* (Nutt.) Gray
§ *Halodes* (Gray) L. Benson
Ranunculus cymbalaria Pursh
SUBGENUS *Batrachium* (DC.) Gray
§ *Batrachium* DC.
Ranunculus trichophyllus Chaix in Vill.
Ranunculus longirostris Godr.

KEY TO THE SPECIES OF Ranunculus IN ILLINOIS

1. Petals white; achenes covered by horizontal wrinkles _____ 2
1. Petals yellow; achenes smooth or variously marked, but not with horizontal wrinkles _____ 3
    2. Leaves becoming limp after removal from the water; beak of achene less than 1 mm long, or absent _____ 1. *R. trichophyllus*
    2. Leaves remaining firm after removal from the water; beak of achene about 1 mm long _____ 2. *R. longirostris*
3. At least some of the leaves simple and unlobed _____ 4
3. None of the leaves simple and unlobed _____ 13
    4. All leaves simple and unlobed _____ 5
    4. At least some of the leaves lobed or divided _____ 8
5. Leaves reniform, cordate; petals often slightly shorter than the

sepals _____ 3. R. cymbalaria
5. Leaves linear to lanceolate, tapering to base; petals equaling or slightly
    longer than the sepals _____ 6
    6. Petals 5–7 in number, 3–9 mm long; stamens 12–50 _____ 7
    6. Petals 1–3 (−5) in number, 1.0–2.5 mm long; stamens 3–10 _____
    _____ 6. R. pusillus
7. Perennial; stamens 25–50; achenes flattened, about 2 mm long _____
    _____ 4. R. ambigens
7. Annual; stamens 12–25; achenes plump, about 1 mm long _____
    _____ 5. R. laxicaulis
    8. Petals longer than the sepals _____ 9
    8. Petals equaling or shorter than the sepals _____ 11
9. Achenes copiously spiny, up to 4.5 mm long, the beak up to 3 mm
    long _____ 25. R. arvensis
9. Achenes smooth, up to 2.5 mm long, the beak less than 1 mm long 10
    10. Stamens in one or two series; fruiting head less than 6 mm thick;
    sepals without long white hairs _____ 7. R. harveyi
    10. Stamens in 3–5 series; fruiting head 6–10 mm thick; sepals with
    long white hairs _____ 8. R. rhomboideus
11. Plants more or less fleshy; achenes with corky thickenings at base;
    stems often hollow _____ 9. R. sceleratus
11. Plants not fleshy; achenes without corky thickenings at base; stems not
    hollow _____ 12
    12. Achenes shiny; receptacle pubescent; roots slender _____
    _____ 10. R. abortivus
    12. Achenes dull; receptacle glabrous, except sometimes near the tip;
    roots thickened _____ 11. R. micranthus
13. Plants aquatic or, if creeping in mud, some of the leaves finely
    dissected _____ 14
13. Plants not truly aquatics; leaves not finely divided _____ 15
    14. Achenes rugose on the sides, corky-thickened at the base; beak of
    achene about 1.5 mm long _____ 12. R. flabellaris
    14. Achenes smooth on the sides, not corky-thickened at the base; beak
    of achene up to 0.8 mm long _____ 13. R. gmelini
15. Petals up to 6 mm long _____ 16
15. Petals 6 mm long or longer _____ 18
    16. Achenes with short, hooked spines; petals 1–2 mm long _____
    _____ 14. R. parviflorus
    16. Achenes smooth; petals 2–6 mm long _____ 17
17. Petals about equaling the sepals; achenes flat, with strongly recurved
    beaks; terminal lobe of leaves not stalked _____ 15. R. recurvatus

17. Petals distinctly shorter than the sepals; achenes not flat, with nearly straight beaks; terminal lobe of leaves stalked   16. *R. pensylvanicus*
    18. Achenes smooth on the sides _____ 19
    18. Achenes papillate on the sides _____ 24. R. sardous
19. Petals at least half as broad as long _____ 20
19. Petals less than half as broad as long _____ 25
    20. Terminal segment of leaves not stalked _____ 17. *R. acris*
    20. Terminal segment of leaves stalked _____ 21
21. Stems erect or ascending, not rooting at the nodes _____ 22
21. Stems creeping and rooting at the nodes, at least at maturity ___ 23
    22. Beak of achene less than 1 mm long, usually curved _____
    _____ 18. *R. bulbosus*
    22. Beak of achene more than 1 mm long, more or less straight _____
    _____ 19. *R. hispidus*
23. Achenes plump, the beak up to 1.5 mm long _____ 22. *R. repens*
23. Achenes flattened, the beak 1.5–3.0 mm long _____ 24
    24. Achenes up to 3.5 (–4.5) mm long, with a low narrow keel near the margin _____ 20. *R. septentrionalis*
    24. Achenes 3.5–5.0 mm long, with a high broad keel near the margin _____ 21. *R. carolinianus*
25. Some of the roots tuberous-thickened _____ 23. *R. fascicularis*
25. All roots fibrous _____ 21. *R. carolinianus*

    1. Ranunculus trichophyllus Chaix in Vill. Hist. Pl. Dauph. 1:335. 1786. *Fig. 14.*
    Ranunculus capillaceus Thuill. Fl. Par. ed 1, 1:278. 1799.
    Ranunculus aquatilis L. var. *capillaceus* (Thuill.) DC. Prodr. 1:26. 1824.
    Batrachium trichophyllum (Chaix) F. Schultz, Arch. Fl. France et All. 1:107. 1848.
    Ranunculus aquatilis L. var. *trichophyllus* (Chaix) Gray, Man. Bot. ed. 5, 40. 1867.

Aquatic perennials, stems submersed, rooting at the lower nodes, much branched, up to 2.5 mm in diameter, generally glabrous; floating leaves absent; submersed leaves repeatedly dissected into filiform segments, becoming limp after removal from the water, to 5 cm long, glabrous, the petiole up to 2 cm long, free from the stipule, the stipule sheathing and broad at the base; flowers borne at the surface of the water, 0.8–1.5 cm across; sepals 5, ovate, acute, spreading, glabrous, 2–4 mm long, about half as long as the petals; petals 5, white, obovate, 4–8 mm long, up to 2.5 mm broad; stamens 10–25; receptacle hirsutulous with tufts of hairs; fruiting head

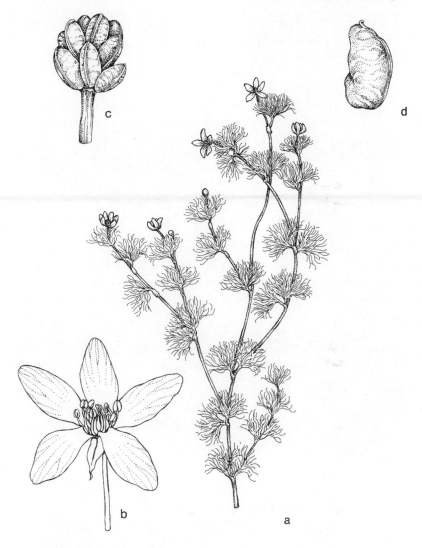

14. *Ranunculus trichophyllus* (White Water-crowfoot). a. Habit, X½. b. Flower, X2½. c. Fruiting head, X6. d. Achene, X15.

globose, to 1 cm in diameter, with (10–) 15–25 achenes; achenes obovoid, 1.0–1.5 mm long, rugose, glabrous or nearly so, the beak up to 0.3 mm long.

COMMON NAME: White Water-crowfoot.

HABITAT: Ponds and slow streams.

RANGE: Labrador to Alaska, south to Baja California (Mexico), New Mexico, South Dakota, central Illinois, and New Jersey.

ILLINOIS DISTRIBUTION: In the northern half of the state; also St. Clair, Saline, and Wabash counties.

This species is best distinguished from *R. longirostris*, the other white-flowered buttercup in Illinois, by its leaves which become limp after their removal from water. The beak of the achene in *R. trichophyllus* is much shorter than the beak in *R. longirostris*.

Ranunculus trichophyllus and *R. longirostris* are the only members of the genus in Illinois belonging to § *Batrachium*. Section *Batrachium* is distinguished by its rugose achenes and its white flowers. Some botanists have placed these species in the segregate genus *Batrachium* (DC.) S. F. Gray.

Benson (1948) and others believe that material assignable to *R. trichophyllus* should not be considered specifically distinct from the Old World *R. aquatilis* L. They would call our plants *R. aquatilis* L. var. *capillaceus* (Thuill.) DC.

Since the Old World *R. aquatilis* possesses floating leaves, petals 1 cm long or longer, about 30 stamens, and achenes 2 mm long, I prefer to give *R. trichophyllus* species status.

2. **Ranunculus longirostris** Godr. Mem. Roy. Soc. Nancy 39. 1839. *Fig. 15.*

Batrachium longirostre (Godr.) F. Schultz, Arch. Fl. France et All. 1:71. 1842.

Ranunculus aquatilis L. var. *longirostris* (Godr.) Laws. Trans. Roy. Soc. Can. 2:45. 1884.

Aquatic perennials; stems floating, rooting at the lower nodes, sparsely branched, up to 3 mm in diameter, generally glabrous; leaves floating or submersed, repeatedly dissected into filiform segments, remaining firm after removal from the water, to 2 cm long, glabrous, the petiole up to 1 cm long, adnate nearly the entire length to the stipule, the stipule sheathing and broad at the base; flowers borne at the surface of the water, 1–2 cm across; sepals 5, elliptic, acute to subacute, spreading, glabrous, 3–4 mm long, about half as long as the petals; petals 5, white, obovate, 4–10 mm long, up to 5 mm broad; stamens 10–20; receptacle hispid; fruiting heads globose, up to 7 mm in diameter, with 7–25 achenes; achenes obovoid, 1.2–1.7 mm long, rugose, glabrous or rarely hispidulous, the beak 0.7–1.2 mm long.

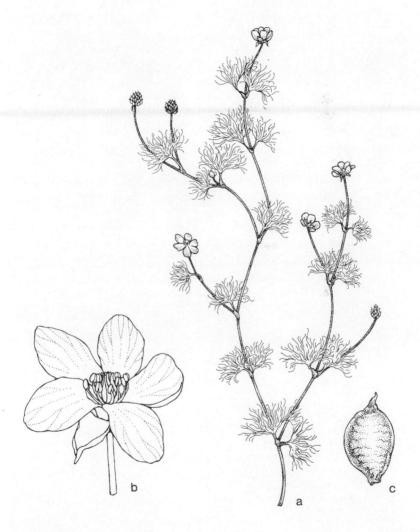

15. *Ranunculus longirostris* (White Water-crowfoot).   *a.* Flowering and fruiting branch, X½.   *b.* Flower, X2½.   *c.* Achene, X15.

COMMON NAME: White Water-crowfoot.

HABITAT: Ponds and slow streams.

RANGE: Quebec to Saskatchewan to Idaho, south to Nevada, Texas, and Alabama.

ILLINOIS DISTRIBUTION: Mostly in the northern counties, extending southward to Franklin, St. Clair, and Wabash counties.

Gleason (1952) believes North American material sufficiently similar to the Old World *R. circinatus* Sibth. to consider them the same.

Several early botanists referred our Illinois plants to *R. divaricatus*, but the true *R. divaricatus* Schrank of the Old World seems to be a different species.

Ranunculus longirostris differs from *R. trichophyllus* by its leaves which remain firm after removal from water, by its longer beak of the achene, and by its petioles adnate to the stipules.

This species flowers from May through August.

3. **Ranunculus cymbalaria** Pursh, Fl. Am. Sept. 392. *Fig. 16.*

Ranunculus cymbalaria Pursh var. *typicus* L. Benson, Am. Midl. Nat. 40:215. 1948.

Tufted perennial with slender stolons; leaves mostly all basal; blades ovate to reniform, cordate, crenate, sometimes with three rounded lobes at apex, glabrous or nearly so, to 25 mm long, nearly as broad; petioles to 5 cm long, glabrous or nearly so; stipules to 1 cm long; flowering scape to 20 cm long, with 1–10 flowers, the flowers 6–10 mm across; sepals 5, elliptic, acute to subacute, spreading, glabrous, to 5 mm long, nearly as long as to slightly longer than the petals; petals usually 5, bright yellow, narrowly obovate, to 5 mm long, to 3 mm broad; stamens 15–25; receptacle pubescent; fruiting head short-cylindric, to 10 (–12) mm long, with 40–150 achenes; achenes oblongoid, 1.3–1.6 mm long, nerved on each face, glabrous, the beak up to 0.3 mm long, straight, or slightly curved.

COMMON NAME: Seaside Crowfoot.

HABITAT: Wet soil.

RANGE: Greenland to Alaska, south to Washington, New Mexico, Texas, Arkansas, and New Jersey; South America; Europe: Asia.

ILLINOIS DISTRIBUTION: Restricted to the northern one-fourth of the state.

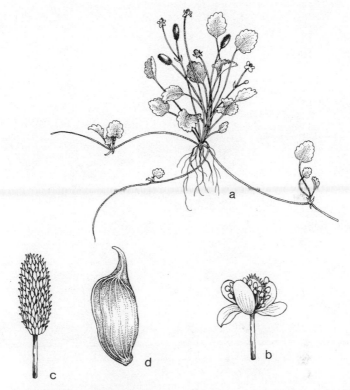

16. Ranunculus cymbalaria (Seaside Crowfoot). *a*. Habit, X¼. *b*. Flower, X2½.
c. Fruiting head, X2. *d*. Achene, X20.

This extremely rare species is the only North
American representative of § *Halodes*, a section distinguished by its simple leaves, veiny achenes, and elongated receptacles.

This is the only buttercup in Illinois with all simple, generally
unlobed leaves and cordate blades.

The flowering time for the seaside crowfoot is May to August.

4. Ranunculus ambigens Wats. Bibl. Ind. N. Am. Bot. 1:16.
1878. *Fig. 17.*

Ranunculus obtusiusculus Raf. Med. Repos. N.Y. II, 5:359.
1808, *nomen confusum.*

Ranunculus flammula L. var. *major* Hook. Fl. Bor. Am. 1:11.
1829.

17. *Ranunculus ambigens* (Spearwort). *a*. Habit, X½. *b*. Flower, X3. *c*. Fruiting head, X3. *d*. Achene, X10.

Perennial; stems rooting at the lower nodes, ascending, sparsely branched except near the inflorescence, hollow, to 75 cm tall, glabrous; basal leaves absent; cauline leaves simple, lanceolate, acuminate, cuneate, remotely denticulate to entire, glabrous, to 12 cm long, to 2 (–3) cm broad, the petioles dilated; flowers few to several in a corymb, 1.2–2.0 cm across; sepals 5, elliptic, acute to subacute, spreading, glabrous, 4–7 mm long, shorter than the petals; petals 5–6, yellow, oblanceolate, 5–8 mm long, up to 3 mm broad; stamens 25–40; receptacle glabrous; fruiting head globose to ovoid, to 7 mm long, with 10–25 achenes; achenes obovoid, compressed, 1.5–2.5 mm long, glabrous, minutely reticulate, the beak 1.0–1.3 mm long, straight.

COMMON NAME: Spearwort.

HABITAT: Swampy woods and ditches.

RANGE: Maine to Minnesota, south to Louisiana and Georgia.

ILLINOIS DISTRIBUTION: Fulton, Jackson, St. Clair, and Wabash counties, but not collected since 1891.

This species has been confused with R. *laxicaulis* in the past, but it differs by its perennial habit, its more numerous stamens, and its larger, compressed achenes.

This species has not been found in Illinois since Schneck's collection from Wabash County in 1891. It flowers from June to September.

September.

5. **Ranunculus laxicaulis** (Torr. & Gray) Darby, Bot. S. States II, 4. 1841. *Fig. 18.*

Ranunculus flammula L. var. *laxicaulis* Torr. & Gray, Fl. N. Am. 1:16. 1838.

Ranunculus texensis Engelm. apud Engelm. & Gray, Bost. Journ. Nat. Hist. 5:210. 1845.

Annual from fibrous roots; stems rooting at the lower nodes, erect to ascending, much branched, to 65 cm tall, glabrous; basal leaves simple, oblong to elliptic to ovate, obtuse, cordate to truncate to cuneate, entire or denticulate, to 4 cm long, to 3 cm broad, glabrous, the petioles to 10 cm long; cauline leaves linear to elliptic to lanceolate, acute to subacute, cuneate, entire or denticulate, to 6 cm long, to 1.2 cm broad, glabrous, sessile or short-petiolate; flowers few to several, 5–15 mm across; sepals 5, ovate, acute, spreading, 2–3 mm long, glabrous or nearly so; petals 5–7, yellow, obovate, 3–9 mm long, up to 2.5 mm broad; stamens 20–25;

18. Ranunculus laxicaulis (Spearwort). *a*. Habit, X½. *c*. Flower, X¼. *d*. Fruiting head, X3. *e*. Achene, X15.

receptacle glabrous; fruiting head hemispherical, up to 5 mm in diameter, with 15–50 achenes; achenes obovoid, plump, 1.0–1.3 mm long, glabrous, the beak 0.1–0.2 mm long.

COMMON NAME: Spearwort.

HABITAT: Wet woods and ditches.

RANGE: Connecticut to Kansas, south to Texas and Florida.

ILLINOIS DISTRIBUTION: Mostly in the southern half of Illinois; also Fulton County.

From Forbes (1870) to Bailey (1949) this species was referred to by almost every Illinois botanist as *R. oblongifolius* Ell. However, *R. oblongifolius* Ell. actually is a synonym for the similar but smaller *R. pusillus* Poir. *Ranunculus texensis* Engelm., which occasionally has been used for our species, is predated by *R. laxicaulis*.

This species shows considerable variation in the number of petals and stamens. Its tiny, plump achenes serve best to distinguish it from *R. ambigens*.

The time for flowering for this species is from May to July.

6. Ranunculus pusillus Poir. ex Lam. Encycl. 6:99. 1804. *Fig. 19.*

Ranunculus oblongifolius Ell. Sketch. 2:58. 1816.

Ranunculus pusillus Poir. var. *typicus* L. Benson, Am. Midl. Nat. 40:197. 1948.

Annual from fibrous roots; stems rooting at the lower nodes, erect or ascending, sparsely branched, to 50 cm tall, glabrous; basal leaves simple, oblong to ovate, acute to subacute, usually rounded at the base, entire or nearly so, to 3 cm long, to 1.5 cm broad, glabrous, the petioles to 6 cm long; cauline leaves linear to lanceolate, acute to subacute, cuneate, entire or sparsely denticulate, to 5 cm long, to 5 mm broad, glabrous, sessile or nearly so; flowers few, to 5 mm across; sepals 5, ovate, acute, spreading, glabrous, 1.0–1.5 mm long; petals 1–3 (–5), yellow, obovate, 1.0–1.5 mm long, about 1 mm broad; stamens 3–10; receptacle glabrous; fruiting head hemispherical, up to 4 mm in diameter, with 15–100 achenes; achenes oblongoid, plump, 0.9–1.1 mm long, glabrous, the beak 0.1–0.2 mm long.

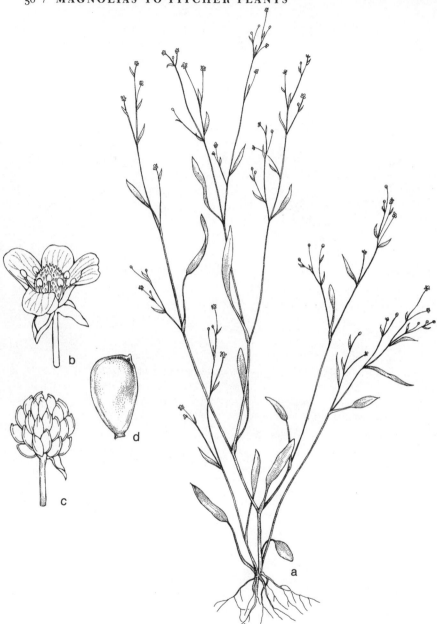

19. Ranunculus pusillus (Small Spearwort).   *a*. Habit, X½.   *b*. Flower, X7½.   *c*. Fruiting head, X5.   *d*. Achene, X20.

COMMON NAME: Small Spearwort.

HABITAT: Swamps, wet woods, ditches.

RANGE: New York to Missouri, south to Texas and Florida; California.

ILLINOIS DISTRIBUTION: Confined to the southern one-third of Illinois.

This usually delicate annual occupies the same habitats as *R. laxicaulis* and sometimes is confused with it. The petals in *R. pusillus* number from one to three, occasionally 5, while the stamens never exceed ten in number.

7. **Ranunculus harveyi** (Gray) Britt. Mem. Torrey Club 5:159. 1894.

Ranunculus abortivus L. var. *harveyi* Gray, Proc. Am. Acad. 21:372. 1886.

Perennial from fibrous roots; stems erect, moderately branched, glabrous or pilose, to 45 cm tall; basal leaves simple, suborbicular to reniform, to 3 cm long, usually a little broader, broadly rounded at the apex, cordate to subcordate at the base, crenate, glabrous or pilose, the petioles to 10 cm long; cauline leaves relatively few, deeply three-parted, the divisions linear-spatulate, obtuse to sub-acute at the apex, long-tapering to the base, to 6 cm long, to 1 cm broad, glabrous or pilose; flowers several, to 1.8 cm across; sepals 5, obovate, subacute, spreading, glabrous, yellow-green, 3.5–4.5 mm long; petals 5–8, yellow, oblong to elliptic, 6–8 mm long, 2–3 mm broad, about twice as long as the sepals; stamens more than 20, in one or two series; receptacle glabrous or nearly so; fruiting head short-cylindrical to ovoid, to 5 mm long, to 4 mm thick, with 20–50 achenes; achenes obovoid to nearly globose, somewhat compressed, 1.3–1.6 mm long, glabrous, the slender beak less than 1 mm long.

Two forms may be distinguished in Illinois.

1. Stems and leaves glabrous _____ 7a. *R. harveyi* f. *harveyi*
1. Stems and leaves pilose _____ 7b. *R. harveyi* f. *pilosus*

7a. Ranunculus harveyi (Gray) Britt. f. **harveyi** *Fig. 20.*

Stems and leaves glabrous.

COMMON NAME: Harvey's Buttercup.
HABITAT: Wooded slopes.
RANGE: Southern Illinois; southern Missouri; Arkansas; western Alabama.
ILLINOIS DISTRIBUTION: Known from Fayette, Jackson, Macoupin, Randolph, and Effingham counties.

Harvey's buttercup exhibits the most restricted overall range of any *Ranunculus* that occurs in Illinois, being known only from four states. In Alabama and Illinois, it is apparently restricted to a few counties in each state.

The first Illinois collection was made by me from Piney Creek ravine in Randolph County on April 24, 1954. I have subsequently found this species along Rock Castle Creek in Randolph County and in Reed's Creek Canyon in Jackson County. At each station the species is only moderately abundant. Since 1970, this species has also been found at Cedar Lake Reservoir in Jackson County and in Effingham, Fayette, and Macoupin counties.

This buttercup resembles *Ranunculus micranthus* or *R. abortivus* in most characters, except that the petals in *R. harveyi* are usually about twice as long as the sepals, while in the other two species, the petals never exceed the sepals.

The flowering time for this species is from mid-April to early May. By late summer, the plants are no longer apparent.

7b. Ranunculus harveyi (Gray) Britt. f. **pilosus** (Benke) Palmer & Steyerm. Ann. Mo. Bot. Gard. 22:540. 1935.

Ranunculus harveyi var. *pilosus* Benke, Rhodora 30:200. 1928.
Stems and leaves pilose.

COMMON NAME: Harvey's Buttercup.
HABITAT: Wooded slopes.
RANGE: Southern Illinois; southern Missouri; Arkansas.
ILLINOIS DISTRIBUTION: Randolph County.

A collection of the hairy form of Harvey's buttercup was made by me from Rock Castle Creek in Randolph County. Except for the pubescence, there appear to be no other differences between this form and the typical form.

20. *Ranunculus harveyi* (Harvey's Buttercup). *a*. Habit, X½. *b*. Flower, X3½.
c. Achene, X17½.

21. Ranunculus rhomboideus (Prairie Buttercup). *a.* Habit, X½. *b.* Flower, X3½.
c. Fruiting head, X2½. *d.* Achene, X7½.

8. Ranunculus rhomboideus Goldie, Edinb. Phil.
Journ. 6:329. 1822. *Fig. 21.*

Ranunculus ovalis Raf. Prec. Dec. 36. 1814, *nomen nudum.*

Perennial from somewhat thickened roots; stems erect, moderately
branched, hirsute or villous (at least below), to 25 cm tall; basal
leaves simple, obovate to oblong to ovate, subacute to rounded at
the apex, cuneate to rounded to subcordate at the base, crenate to
dentate except at the entire base, hirsute to villous to nearly glabrous, to 5 cm long, nearly as broad, the petioles to 8 cm long,

pilose; cauline leaves simple or usually 3– to 7–cleft, sessile, the divisions linear to linear-lanceolate, obtuse to subacute, cuneate, hirsute to villous to nearly glabrous; flowers few, to 2.5 cm across, on villous pedicels; sepals 5, narrowly obovate, subacute, spreading, with long white hairs, yellow-green, 5–6 mm long, about half as broad; petals 5, yellow, narrowly obovate, 6–9 mm long, about 1½ times as long as the sepals; receptacle with long pubescence; stamens 25–50, in 3–5 series; fruiting head globose, 5–10 mm in diameter, with 30 or more achenes; achenes obovoid, flattened below, becoming plump above, 1.8–2.5 mm long, glabrous, the slender beak less than 0.5 mm long.

COMMON NAME: Prairie Buttercup.
HABITAT: Prairies.
RANGE: Ontario to British Columbia, south to Nebraska, Illinois, and New York; Quebec.
ILLINOIS DISTRIBUTION: Restricted to the extreme northern counties.

The prairie buttercup is one of the rarest species in Illinois today, although earlier collections indicate that it was more plentiful in the past.

This species is distinguished by several characteristics: unlobed basal leaves, pubescence of stems and leaves, and petals longer than sepals. Swink (1974) states that this species grows on well-drained morainal hills.

Prairie buttercup flowers during May.

9. Ranunculus sceleratus L. Sp. Pl. 551. 1753. *Fig. 22.*

Rather fleshy annual with fleshy, fibrous roots; stems erect, seldom rooting at the lower nodes, much branched, hollow, glabrous, to 75 cm tall; basal leaves simple, somewhat succulent, reniform in outline, deeply 3-cleft, the divisions often further divided, the ultimate lobes broadly rounded, the blades cordate at the base, glabrous, to 4 cm long, nearly as broad or broader, the petioles to 15 cm long, glabrous; cauline leaves smaller, usually cleft into three linear-oblong divisions; flowers numerous, to 1 cm across; sepals 5, ovate, acute, spreading, pilose or glabrate, yellow-green, 2–4 mm long, over half as broad; petals 5, pale yellow, obovate, 1.5–3.5 mm long, slightly shorter than the sepals; stamens 10–25; receptacle glabrous or puberulent; fruiting head globose to cylindric, to 1 cm long, to 6 mm thick, with 30 or more achenes; achenes obovoid, plump,

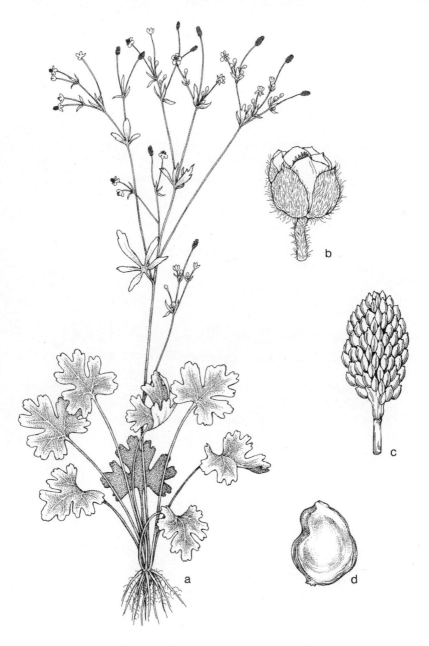

22. *Ranunculus sceleratus* (Cursed Crowfoot). *a*. Habit, X⅓. *b*. Flower, X2½. *c*. Fruiting head, X3½. *d*. Achene, X20.

0.8–1.2 mm long, the surface sometimes with microscopic depressions, glabrous, the minute beak about 0.1 mm long.

COMMON NAME: Cursed Crowfoot.

HABITAT: Wet meadows; ditches; river banks.

RANGE: Newfoundland to Alaska, south to California and Florida; Europe; Asia.

ILLINOIS DISTRIBUTION: Occasional throughout Illinois, although less common in the southern counties.

This buttercup is the most succulent species of *Ranunculus* in the state. The lower leaves are fleshy, and the hollow stem sometimes measures up to 2 cm in diameter at the base.

The pattern of leaf-cutting is very distinctive for this species.

Although *R. sceleratus* generally grows in very wet or muddy soil, I have observed it in standing water.

This species flowers from May to August.

10. Ranunculus abortivus L. Sp. Pl. 551. 1753.

Perennial herb from slender, unthickened fibrous roots; stems erect, moderately branched, glabrous or pilosulous, to 75 cm tall; basal leaves simple or sometimes 3-cleft, the simple ones ovate to suborbicular, mostly rounded at the apex, cordate to subcordate at the base, crenate, glabrous or pilosulous, to 5 cm long, often nearly as broad or even slightly broader, the 3-cleft leaves usually tapering to the base, the divisions mostly oblanceolate to obovate, the petioles of both kinds of leaves glabrous or puberulent, to 10 cm long; cauline leaves simple or 3-cleft, elliptic to oblanceolate, sessile or on petioles up to 3 cm long; flowers several to many, to 7 mm across; sepals 5, elliptic, subacute to acute, spreading to reflexed, glabrous, yellow-green, 3–4 mm long; petals 5, yellow, usually lustrous, oblong to oval, 2.0–3.5 mm long, more than half as broad, shorter than the sepals; stamens 15–20; receptacle pubescent; fruiting head ellipsoid to short-ovoid, to 7 mm long, to 4 mm thick, with 10–35 achenes; achenes obovoid, plump, 1.4–1.7 mm long, glabrous, lustrous, the slender beak about 0.2 mm long.

Two forms may be distinguished in Illinois.

1. Stems and leaves glabrous _____ 10a. *R. abortivus* f. *abortivus*
1. Stems and leaves pilosulous _____ 10b. *R. abortivus* f. *acrolasius*

10a. Ranunculus abortivus L. f. **abortivus** *Fig. 23.*
Stems and leaves glabrous.

COMMON NAME: Small-flowered Crowfoot.
HABITAT: Fields; moist woods.
RANGE: Maine to Saskatchewan, south to Oklahoma and Florida.
ILLINOIS DISTRIBUTION: Common throughout the state.

This plant occupies a wide variety of habitats from undisturbed to disturbed areas. It is a common weed of lawns.

Distinction of *R. abortivus* from *R. micranthus* is best made by the pubescent receptacle, lustrous achenes, and slender roots of *R. abortivus.*

Fassett reported the New England variety *eucyclus* Fern. from Kane County, but this is in error for the typical variety.

Flowering time in Illinois for this small-flowered crowfoot is from early April to late June.

10b. Ranunculus abortivus L. f. **acrolasius** Fern. Rhodora
40:418. 1938.
Stems and leaves pilosulous.

COMMON NAME: Small-flowered Crowfoot.
HABITAT: Moist woods; fields.
RANGE: Labrador to British Columbia, south to Colorado, South Dakota, Illinois, and Pennsylvania; Alaska.
ILLINOIS DISTRIBUTION: Rare but scattered throughout the state.

The puberulent form of *R. abortivus* is distressingly similar to some material of *R. micranthus* which may be sparsely villous. In those instances, examination of the receptacle is essential for positive identification.

11. Ranunculus micranthus Nutt. ex Torr. & Gray, Fl. N. Am.
1:18. 1838. *Fig. 24.*
Ranunculus abortivus L. var. *micranthus* (Nutt.) Gray, Man. ed.
2, 9. 1856.
Ranunculus delitescens Greene, Am. Midl. Nat. 3:333. 1914.
Ranunculus micranthus var. *delitescens* (Greene) Fern. Rhodora
41:543. 1939.

23. *Ranunculus abortivus* (Small-flowered Crowfoot). *a*. Habit, X½. *b*. Flower, X5. *c*. Fruiting head, X5. *d*. Achene, X12½.

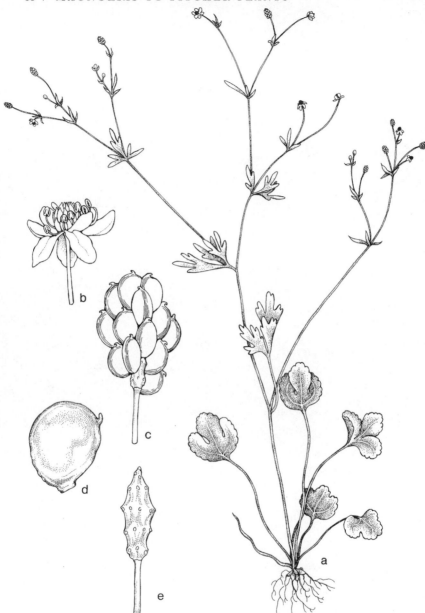

24. *Ranunculus micranthus* (Small-flowered Crowfoot). *a*. Habit, X½. *b*. Flower, X4. *c*. Fruiting head, X7½. *d*. Achene, X15. *e*. Receptacle, X7½.

Perennial from both thickened and slender fibrous roots; stems erect, moderately branched, more or less villous, to 50 cm tall; basal leaves simple or sometimes 3-cleft, the simple ones ovate to suborbicular, mostly rounded at the apex, cordate to subcordate to cuneate at the base, crenate to somewhat dentate, sparsely villous, to 4 cm long, often nearly as broad, the 3-cleft leaves usually tapering to base, the divisions mostly oblanceolate to obovate, the petioles of both kinds of leaves villous, to 10 cm long; cauline leaves simple or 3-cleft, oblanceolate, nearly sessile; flowers relatively few, to 7 mm across; sepals 5, obovate, acute, spreading to reflexed, glabrous or puberulent, yellow-green, 2.5–3.5 mm long; petals 5, yellow, lustrous, linear to oblong, 2–3 mm long, about half as broad, shorter than the sepals; stamens 15–25; receptacle glabrous; fruiting head ellipsoid to short-cylindrical, to 6 mm long, to 4 mm thick, with 10–50 achenes; achenes obovoid, plump, 1.3–1.5 mm long, glabrous, dull, the slender beak less than 0.5 mm long.

COMMON NAME: Small-flowered Crowfoot.
HABITAT: Moist or dry woods.
RANGE: Massachusetts to Missouri, south to Arkansas and Georgia; Oklahoma; South Dakota.
ILLINOIS DISTRIBUTION: Occasional in the southern half of Illinois.

This small-flowered species closely resembles R. abortivus, but differs by the presence of thickened roots, dull achenes, and glabrous receptacles. The villous pubescence on the stems, often suggested as a good field character, may also be found in R. abortivus var. acrolasius.

Variety delitescens (Greene) Fern. (or Ranunculus delitescens Greene) does not exhibit enough clear-cut differences to merit recognition. Fernald (1938) claims that this variety is generally paler and more slender and has basal leaves cuneate to rounded with fewer teeth per margin than in the typical variety. I frankly can see no uniform correlation of these characters. Plants which I have examined which have cuneate leaf bases and fewer teeth are not necessarily pale and slender, while some pale specimens definitely are subcordate at base and many-toothed.

This species flowers from late March to mid-May.

12. **Ranunculus flabellaris** Raf. apud Bigel. Am. Monthly Mag. 2:344. March. 1818. *Fig. 25*.

Ranunculus multifidus Pursh, Fl. Am. Sept. 2:736. 1814, non Forsk. (1775).

Ranunculus delphinifolius Torr. in Eat. Man. Bot. ed. 2, 395. May. 1818.

Ranunculus multifidus var. *terrestris* Gray, Man. ed. 5, 41. 1867.

Ranunculus delphinifolius f. *rosiflorus* Clute, Am. Bot. 34:106. 1928.

Submersed or emergent aquatic perennials; stems hollow, floating or reclining, rooting at the lower nodes, much branched; submersed leaves ternately compound, to 10 cm long, about as broad, the linear divisions 1.0–1.5 mm broad, glabrous; emersed leaves (when present) three-cleft, with each division 3-parted, glabrous; flowers several, to 3 cm across, on stout pedicels up to 6 cm long; sepals 5, ovate, acute, spreading, yellow-green, glabrous or nearly so, 5–8 mm long, about half as long as the petals; petals 5, yellow or rarely roseate, obovate, 7–14 mm long; stamens 50 or more; receptacle pubescent; fruiting head subglobose, to 1.5 cm in diameter, with 50 or more achenes; achenes obovoid, plump, 1.5–2.5 mm long, rugose on the sides, with the margin corky-thickened, glabrous, the broad, flat beak about 1.5 mm long.

COMMON NAME: Yellow Water-crowfoot.

HABITAT: Swamps and ponds and quiet pools.

RANGE: Maine to British Columbia, south to California, Louisiana, and North Carolina.

ILLINOIS DISTRIBUTION: Occasional throughout the state.

This species usually grows in standing water where all its leaves are submersed and its flowers float on the surface of the water atop thickened pedicels. Occasionally plants may be found stranded on land. If these terrestrial forms persist, they usually develop less intricately divided leaves. Such obvious ecological forms are scarcely worth recognition in a different taxonomic rank.

The flowers are a waxy, golden-yellow, although a roseate form was described by Clute from Braidwood in Will County. I have observed that petals on dried material sometimes have a faint rosy tint.

This species flowers from April to mid-July.

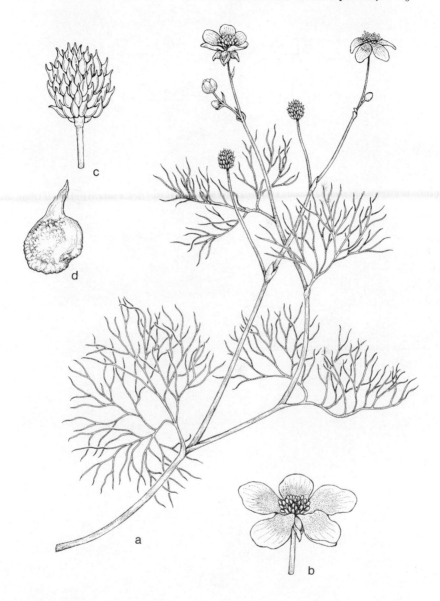

25. Ranunculus flabellaris (Yellow Water-crowfoot). *a*. Habit, X½. *b*. Flower, X1½. *c*. Fruiting head, X2. *d*. Achene, X10.

26. Ranunculus gmelinii var. *hookeri* (Small Yellow Water-crowfoot). *a*. Habit, X½. *b*. Flower, X3. *c*. Achene, X15.

13. **Ranunculus gmelinii** DC. var. **hookeri** (D. Don) L. Benson, Am. Midl. Nat. 40:209. 1948. *Fig. 26*.

Ranunculus purshii Richards. Bot. App. Frankl. 1st Journey 751. 1823.
Ranunculus purshii var. *hookeri* D. Don in G. Don, Gen. Syst. Gard. 1:33. 1831.

Submersed or emergent aquatic perennials; stems hollow, floating or reclining, rooting at the lower nodes, much branched; submersed leaves either deeply dissected in many linear divisions or 3– to 5–cleft into obovate, cuneate lobes, glabrous or pilose, to 10 cm long, about as broad; emersed leaves (when present) three-cleft, with each division 3-parted, glabrous or pilose; flowers 1–4, to 1.8 cm across, on moderately stout pedicels up to 3 cm long; sepals 5, suborbicular to ovate, rounded or subacute at the apex, spreading, yellow-green, 3–5 mm long, half to nearly as long as the petals; petals 5, yellow, obovate, 4–7 mm long; stamens 20–40; receptacle pubescent; fruiting head subglobose, to 6 mm in diameter, with 50 or more achenes; achenes obovoid, somewhat flattened, 1.0–1.5 mm long, without corky-thickened margins, glabrous, the broad, flat beak 0.4–0.8 mm long.

COMMON NAME: Small Yellow Water-crowfoot.
HABITAT: Ponds.
RANGE: Newfoundland to Alaska, south to Oregon, New Mexico, Illinois, and Maine.
ILLINOIS DISTRIBUTION: Known only from a single collection (Cook Co.: Chicago, *Munroe*, in 1877).

The small yellow-flowered crowfoot is smaller in all respects than *R. flabellaris*. In addition, most of the leaves of *R. gmelinii* var. *hookeri* are divided into broader segments than those of *R. flabellaris*.

The binomial *R. purshii* Richards. is the correct one for this taxon at the rank of species.

Since this plant has not been collected in Illinois since 1877, it probably is extinct in this state.

14. **Ranunculus parviflorus** L. Sp. Pl. ed. 2, 780. 1763. *Fig. 27*.
Annual from slender, fibrous roots; stems erect, much branched, densely villous, to 40 cm tall; basal leaves simple, reniform, usually deeply cordate, 3-parted with each division divided again, the lobes acute, villous, the petioles to 10 cm long, pilose; cauline leaves

27. *Ranunculus parviflorus* (Small-flowered Crowfoot). *a*. Habit, X½. *b*. Flower, X7½. *c*. Fruiting head, X4. *d*. Achene, X17½.

similar to the basal leaves but smaller; flowers few to several, to 4 mm across, on peduncles ultimately longer than the subtending bracts; sepals 5, narrowly ovate, acute, spreading, 0.8–1.2 mm long; petals 5, yellow, narrowly elliptic, barely longer than the sepals; stamens usually 10; receptacle glabrous; fruiting head globose, to 5 mm in diameter, with 10–20 achenes; achenes obovoid, somewhat flattened, 1.3–1.5 mm long, the sides covered with short, broad-based, hooked spines, the margin conspicuous but narrow, the beak triangular, strongly curved, flattened, 0.4–0.6 mm long.

COMMON NAME: Small-flowered Crowfoot.

HABITAT: Field (in Illinois).

RANGE: Native of Europe; naturalized from New York to Missouri, south to Texas and Florida; California; Bermuda; Jamaica; Haiti.

ILLINOIS DISTRIBUTION: Known only from Jackson County (field, Saltpeter Cave, seven miles south of Murphysboro, May 13, 1954, *R. H. Mohlenbrock 2437*).

This species is readily distinguished by its achenes which are densely beset with hooked, broad-based prickles on the sides. The flowers, in addition, are extremely tiny. The somewhat similar-looking *R. pensylvanicus* has smooth achenes and slightly longer petals.

Ranunculus parviflorus flowers in May and June.

15. **Ranunculus recurvatus** Poir. in Lam. Encyc. Meth. 6:125. 1804. *Fig. 28*.

Perennial from slender, fibrous roots; stems erect, often somewhat swollen at the base, branched, villous-hirsute, to 75 cm tall; basal leaves simple, cordate, shallowly to deeply 3-cleft, with each division shallowly lobed or crenate, more or less hirsutulous or pilose above and below, to 7 cm long, nearly as broad or slightly broader, the petioles to 20 cm long, villous-hirsute; cauline leaves similar to the basal leaves but smaller; flowers few to several, to 1 cm across; sepals 5, ovate, acute, reflexed, green, pilose, 3–6 mm long; petals 5, pale yellow, narrowly elliptic, 2–5 mm long, usually slightly shorter than the sepals; stamens 10–25; receptacle hispid; fruiting head globose to ovoid, 5–7 mm in diameter, with 10–25 achenes; achenes obovoid, flattened, 1.5–2.0 mm long (excluding the beak),

28. *Ranunculus recurvatus* (Rough Crowfoot). *a*. Habit, X½. *b*. Flower, X3. *c*. Fruiting head, X3½. *d*. Achene, X12½.

glabrous, minutely pitted, the margin sharply defined, the beak strongly hooked, 1.0–1.5 mm long.

COMMON NAME: Rough Crowfoot.
HABITAT: Wet woods; calcareous fens.
RANGE: Newfoundland to Manitoba, south to Texas and Florida.
ILLINOIS DISTRIBUTION: Occasional throughout the state.

The rough crowfoot is recognized by its dense hairiness, its small, pale yellow petals, its broadly lobed leaves, and its achenes with hooked beaks. The somewhat similar *R. pensylvanicus* has more deeply cleft leaves and shorter, straight beaks on the achenes.

This species occupies wet woodland habitats. It flowers from mid-April into early June.

16. Ranunculus pensylvanicus L. f. Suppl. 272. 1781. *Fig. 29.*

Annual or sometimes perennial herbs from slender, fibrous roots; stems erect, not swollen at the base, hollow, villous-hirsute, somewhat branched, to nearly 1 m tall; basal leaves usually absent at flowering time, deeply 5– to 7–parted, borne on hirsute petioles to 15 cm long; cauline leaves deeply 3-cleft, with each division 3-lobed, at least the terminal division stalked, spreading hirsute; flowers few to several, to 8 mm across; sepals 5, narrowly elliptic, acute to subacute, reflexed, yellow-green, hirsutulous, 3–5 mm long, somewhat longer than the petals; petals 5, pale yellow, oblong to obovate, 2–4 mm long; stamens 15–20; receptacle hirsute; fruiting head oblongoid to cylindric, to 1.5 cm long, to 1 cm across, with 60 or more achenes; achenes obovoid, not conspicuously flattened, 2.0–2.5 mm long, glabrous, with prominent narrow margins, the beak nearly straight, up to 1 mm long.

COMMON NAME: Bristly Crowfoot.
HABITAT: Wet ground, often in ditches.
RANGE: Labrador to Alaska, south to Oregon, Nebraska, Illinois, and West Virginia.
ILLINOIS DISTRIBUTION: Confined to the northern part of Illinois, extending southward as far as St. Clair County; also in Johnson County.

The species most nearly related to *R. pensylvanicus*

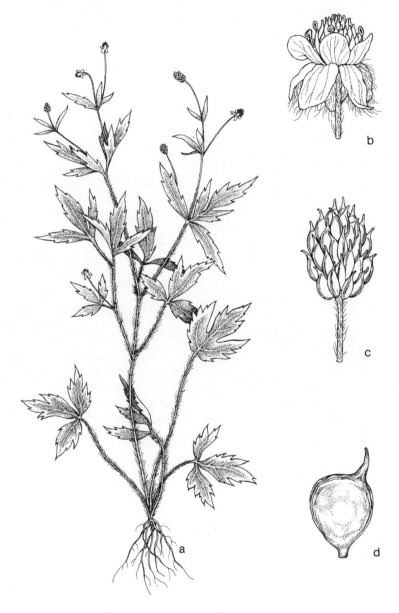

29. *Ranunculus pensylvanicus* (Bristly Crowfoot).   *a*. Habit, X¼.   *b*. Flower, X4. *c*.
Fruiting head, X7½.   *d*. Achene, X10.

apparently is *R. recurvatus*. This latter species is distinguished, however, by its very recurved beaks of the achenes and by its less deeply divided leaves.

The basal leaves of *R. pensylvanicus*, which are divided into more lobes than are the cauline leaves, usually are withered by the time of flowering.

Flowering time for this species is from July until early September.

17. Ranunculus acris L. Sp. Pl. 554. *Fig. 30.*

Perennial herbs from slender, fibrous roots; stems slender or sometimes stout, erect, much branched above, hirsute or nearly glabrous, to nearly 1 m tall; basal leaves deeply 3-parted, each division lobed or deeply toothed, the terminal segment not stalked, appressed-pubescent, borne on hirsute petioles up to 30 cm long; cauline leaves deeply 3- to 5-lobed, becoming progressively shorter petiolate toward the apex of the stem, appressed-pubescent; flowers several, up to 2.5 cm across, on pedicels up to 5 (–10) cm long; sepals 5, narrowly ovate, acute, spreading, greenish, densely pilose, 4–7 mm long; petals 5, bright yellow, obovate, 8–15 mm long, at least half as broad or broader, usually at least twice as long as the sepals; stamens very numerous; receptacle glabrous; fruiting head globose, 5–8 mm in diameter, with 25 or more achenes; achenes obovoid, flattened, 2–3 mm long, glabrous, with a prominent narrow margin, the beak slightly curved, up to 0.5 mm long.

COMMON NAME: Tall Buttercup.

HABITAT: Roadsides, fields, and pastures.

RANGE: Native of Europe; adventive from Newfoundland to Alaska, south to California, Kansas, and Georgia.

ILLINOIS DISTRIBUTION: Occasional in the northern half of the state; also in Jackson County.

The common name of this plant is testimony to the stature that this species may attain. Specimens nearly one meter tall have been seen in Illinois.

The degree of leaf-cutting is variable for this species, as well as the amount of pubescence present on the stems.

The tall buttercup flowers from May to August.

30. *Ranunculus acris* (Tall Buttercup).   *a*. Basal portion of plant, X½.   *b*. Upper portion of plant, X½.   *c*. Flower and buds, X1.   *d*. Petal, X2½.   *e*. Achene, X12½.

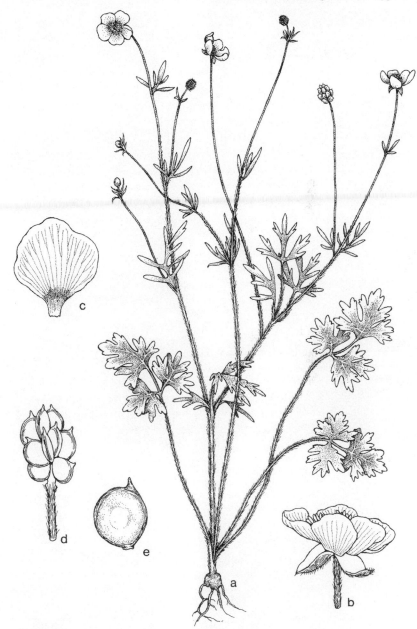

31. Ranunculus bulbosus (Bulbous Buttercup). *a.* Habit, X½. *b.* Flower, X1½. *c.* Petal, X2½. *d.* Fruiting head, X2. *e.* Achene, X5.

18. Ranunculus bulbosus L. Sp. Pl. 554. 1753. *Fig. 31*.

Perennial herbs from a swollen, bulbous base; stems usually stout, erect, much branched, hirsute, to 65 cm tall; basal leaves usually trifoliolate, each leaflet lobed or toothed, the terminal segment stalked, densely or sparsely strigose, borne on hirsute to strigose pedicels up to 30 cm long; cauline leaves deeply 3-lobed or trifoliolate, the uppermost sessile or nearly so; flowers few to several, up to 2.5 cm across, on strigose pedicels up to 10 cm long; sepals 5, ovate, acute, reflexed, greenish-yellow, densely pilose, 4–7 mm long; petals 5, bright yellow, ovate to orbicular, 7–15 mm long, over half as wide, usually about twice as long as the sepals; stamens over 25; receptacle pubescent; fruiting head globose, 6–9 mm in diameter, with 12 or more achenes; achenes obovoid, flattened, 2.5–3.5 mm long, glabrous, with a prominent narrow margin, the beak stout, recurved, 0.5–1.0 mm long.

COMMON NAME: Bulbous Buttercup.

HABITAT: Waste ground.

RANGE: Native of Europe; scattered throughout North America as an adventive.

ILLINOIS DISTRIBUTION: Known from Cook, Henderson, Jackson, and Johnson counties. The first Illinois collection was made in 1875 by H. N. Patterson from Henderson County.

The bulbous buttercup is readily distinguished by its swollen-bulbous base. Several specimens from Illinois have been identified as *R. bulbosus*, but in reality they are mostly *R. hispidus* with a hardened, but scarcely swollen, base. The shorter, recurved back of the achene also distinguishes *R. bulbosus* from *R. hispidus*.

This species flowers from late April to mid-July.

19. Ranunculus hispidus Michx. Fl. Bor. Am. 1:321. 1803.

Perennial herbs occasionally with a thickened swelling at base but scarcely bulbous, with somewhat fleshy roots; stems erect, sometimes branched, hirsute or appressed-pubescent, to 50 cm tall; basal leaves simple and unlobed or deeply 3-parted or even trifoliolate, the terminal segment stalked, hirsute or appressed-pubescent, borne on hirsute or appressed-pubescent petioles up to 20 cm long; cauline leaves usually deeply 3-parted, the uppermost ones sessile or nearly so; flowers 1–6 per stem, up to 2 cm across, on strigose or hirsutulous pedicels up to 10 cm long; sepals 5, ovate, acute, spreading, greenish-yellow, pilose, 4–7 mm long; petals 5, bright yellow, obovate, 7–15 mm long, over half as wide, usually about twice as long as the sepals; stamens usually 40 or more; receptacle hispidulous; fruiting head globose, 6–10 mm in diameter, with 15 or more achenes; achenes obovoid, flattened, 2.0–3.5 mm long, glabrous, with a prominent narrow margin, the beak straight, 1.2–2.0 mm long.

Two varieties may be distinguished in Illinois.

1. Pubescence spreading; achenes 3.0–3.5 mm long (excluding the beak) _____ 19a. *R. hispidus* var. *hispidus*
1. Pubescence appressed; achenes 2.0–2.5 mm long (excluding the beak) _____ 19b. *R. hispidus* var. *marilandicus*

19a. Ranunculus hispidus Michx. var. **hispidus** *Fig. 32.*

Ranunculus repens L. var. *hispidus* (Michx.) Chapm. Fl. So. U.S. 8. 1860.

Pubescence spreading; achenes 3.0–3.5 mm long (excluding the beak).

COMMON NAME: Bristly Buttercup.

HABITAT: Woods.

RANGE: New York to Missouri, south to Arkansas and Georgia.

ILLINOIS DISTRIBUTION: Occasional in the southern three-fifths of Illinois, less common elsewhere.

This variety, with spreading pubescence, is the common type of *R. hispidus* in Illinois. It is often readily confused, however, with *R. fascicularis, R. septentrionalis* var. *caricetorum,* and *R. bulbosus.*

In *R. fascicularis,* the roots are tuberous-thickened and the petals are usually less than half as broad as long. Without having seen the tuberous-thickened roots of *R. fascicularis,* one may think the fleshy roots of *R. hispidus* are tuberous-thickened.

Likewise, specimens of *R. hispidus* var. *hispidus* with thickened, swollen bases may be mistaken for *R. bulbosus* until the bulbous base of *R. bulbosus* is actually seen. To distinguish these two, examination of the beak of the achene should be made. *Ranunculus hispidus* has a nearly straight beak, while *R. bulbosus* has a recurved beak.

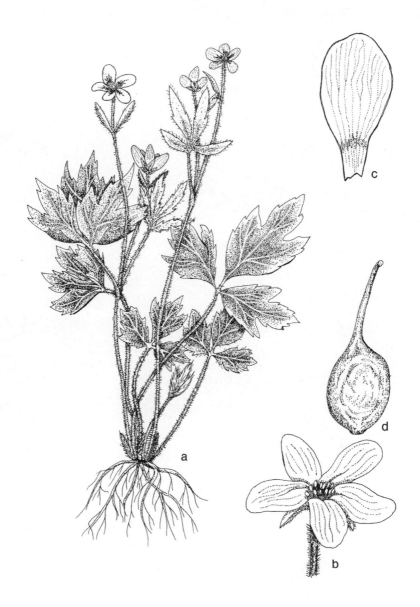

32. Ranunculus hispidus (Bristly Buttercup). *a.* Habit, X½. *b.* Flower, X1½. *c.* Petal, X3. *d.* Achene, X10.

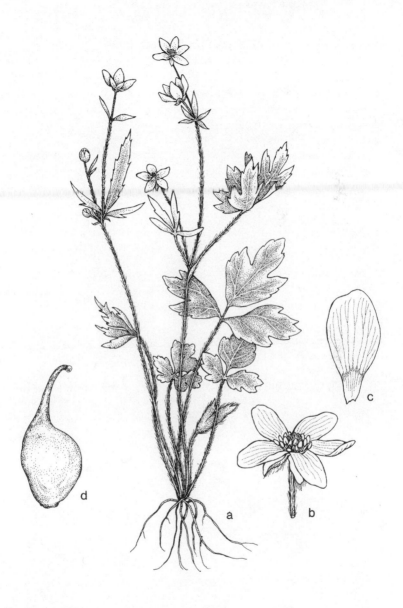

33. *Ranunculus hispidus* var. *marilandicus* (Buttercup). *a*. Habit, X½. *b*. Flower, X1½. *c*. Petal, X3. *d*. Achene, X10.

Ranunculus hispidus var. *hispidus* is similar in appearance to the hirsute *R. septentrionalis* var. *caricetorum*, but differs by its absence of creeping stems which root at the nodes.

Ranunculus hispidus var. *hispidus* is one of the first woodland buttercups to flower, coming into bloom about the second week in April in the southernmost counties of Illinois.

Patterson (1876) referred to this plant in Illinois as *R. repens* L. var. *hispidus* (Michx.) Chapm.

19b. Ranunculus hispidus Michx. var. **marilandicus** (Poir.) L. Benson, Am. Midl. Natl. 40(1):84. 1948. *Fig.* 33.

Ranunculus marilandicus Poir. in Lam. Encyc. Meth. 6:126. 1804.

Ranunculus septentrionalis Poir. var. *marilandicus* (Poir.) Torr. & Gray, Fl. N. Am. 1:21. 1838.

Ranunculus hispidus var. *falsus* Fern. Rhodora 22:30. 1920.

Pubescence appressed; achenes 2.0–2.5 mm long (excluding the beak).

COMMON NAME: Buttercup.

HABITAT: Dry woodlands.

RANGE: Massachusetts to Wisconsin and Iowa, south to Arkansas and South Carolina.

ILLINOIS DISTRIBUTION: Rare in southern Illinois.

Correlating with the appressed-pubescence of var. *marilandicus* is the short achene which is only 2.0–2.5 mm long.

The appressed-pubescent *R. hispidus* var. *marilandicus* is much less common in Illinois than the hirsute var. *hispidus*. It appears at first glance, on the basis of its pubescence, to be *R. septentrionalis*. However, this latter species creeps and roots at the nodes, while *R. hispidus* var. *marilandicus* is erect and nonrooting at the nodes.

Fernald's var. *falsus* is apparently synonymous with var. *marilandicus*.

This variety blooms in April and May.

20. Ranunculus septentrionalis Poir. in Lam. Encyc. Meth. 6:125. 1803.

Perennial herbs with prominent stolons creeping and rooting at the nodes; stems trailing to ascending, branched, sometimes hollow, glabrous or nearly so to appressed-pubescent to densely retrorse-hirsute, to 85 cm long; all leaves similar, deeply 3-parted or trifoliolate, the terminal segment stalked, appressed-pubescent to hirsute, borne on appressed-pubescent to retrorse-hirsute petioles up to 30 cm long, the uppermost leaves often shorter petiolate; flowers 1–10 per stem, up to 2.5 cm across, on appressed- or spreading-hispid pedicels to 6 cm long; sepals 5, narrowly ovate, acute, spreading, greenish-yellow, nearly glabrous to pilose to hispidulous, 6–10 mm long; petals 5, bright yellow, mostly obovate, 8–15 mm long, usually over half as wide, usually less than twice as long as the sepals; stamens usually 40 or more; receptacle hispidulous; fruiting head globose to ovoid, 8–14 mm long, nearly as thick, with 15–30 achenes; achenes obovoid, flattish, to 3.5 mm long, glabrous, with a low narrow keel near the margin, the beak straight or curved, 1.8–3.0 mm long.

Two varieties may be distinguished in Illinois.

1. Petioles and lower part of stems glabrous or appressed-pubescent. _____ 20a. *R. septentrionalis* var. *septentrionalis*
1. Petioles and lower part of stems with retrorse pubescence _____ _____ 20b. *R. septentrionalis* var. *caricetorum*

20a. Ranunculus septentrionalis Poir. var. **septentrionalis** *Fig. 34.*

Petioles and lower part of stems glabrous or appressed-pubescent.

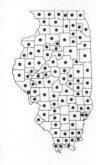

COMMON NAME: Swamp Buttercup.

HABITAT: Low woods, ditches, floodplains, and swamps.

RANGE: Labrador to Manitoba, south to Texas and Georgia.

ILLINOIS DISTRIBUTION: Common throughout the state.

The typical variety of *R. septentrionalis* is not conspicuously pubescent and therefore is readily distinguished from hairy forms of *R. hispidus* and *R. fascicularis*. There is close similarity, however, between var. *septentrionalis* and *R. hispidus* var. *marilandicus*.

The creeping stems of *R. septentrionalis* are distinctive, although this character often is not apparent in most herbarium specimens. Sparsely pubescent forms of *R. fascicularis* are distinguished by their tuberous-thickened roots.

34. Ranunculus septentrionalis (Swamp Buttercup). *a*. Habit, X½. *b*. Flower,
X1½. *c*. Petal, X3. *d*. Fruiting head, X2½. *e*. Achene, X5.

Until the last decade of the nineteenth century, Illinois botanists called this plant *Ranunculus repens* L. *Ranunculus repens*, however, is a totally different species.

Ranunculus carolinianus is a similar species, differing by its achene which is larger and which has a high broad keel near the margin.

This variety flowers from late March through June. It sometimes has a brief second flowering period in September or early October.

20b. Ranunculus septentrionalis Poir. var. **caricetorum** (Greene) Fern. Rhodora 38:177. 1936. *Fig.* 35.

Ranunculus caricetorum Greene, Pittonia 5:194. 1903.

Petioles and lower part of stems with retrorse pubescence.

COMMON NAME: Swamp Buttercup.

HABITAT: Low woods and wet areas.

RANGE: Maryland to Minnesota, south to Iowa and Ohio.

ILLINOIS DISTRIBUTION: DeKalb Co.: along Chicago and Northwestern Railroad south of Turk's Grocery, June 12, 1966, *F. A. Swink 73–H*; also DuPage County.

This densely pubescent variety is geographically confined mostly to the midwestern United States. The resemblance to *R. hispidus* var. *hispidus* is great, but *R. septentrionalis* var. *caricetorum* usually has creeping stems.

21. Ranunculus carolinianus DC. Syst. 1:292. 1818. *Fig.* 36.

Ranunculus septentrionalis Poir. var. *pterocarpus* L. Benson, Bull. Torrey Club 68:486. 1941.

Perennial herbs with prominent stolons creeping and rooting at the nodes; stems trailing to ascending, branched, generally appressed-pubescent to nearly glabrous, to 75 cm long; all leaves similar, deeply 3-parted or trifoliolate, the terminal segment glabrous or appressed-pubescent, borne on appressed-pubescent to glabrous petioles up to 25 cm long, the uppermost leaves becoming progressively shorter petiolate; flowers 1–10 per stem, up to 2.5 cm across, on appressed-pubescent or glabrous pedicels to 5 cm long; sepals 5, narrowly ovate, acute, spreading or reflexed, greenish-yellow, glabrous or puberulent, 3.5–5.0 mm long; petals 5, bright yellow, ob-

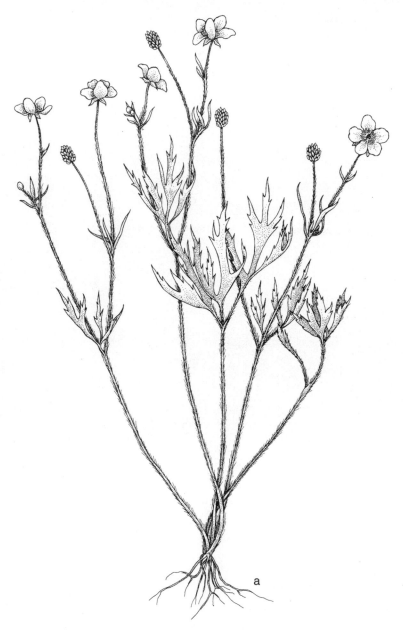

35. *Ranunculus septentrionalis* var. *caricetorum* (Swamp Buttercup). *a*. Habit, X½.

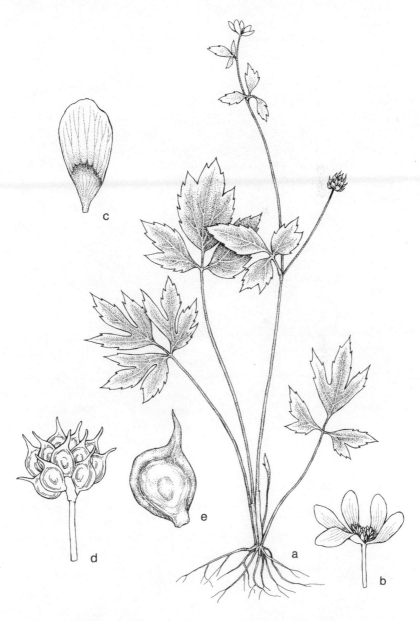

36. Ranunculus carolinianus (Buttercup). *a*. Habit, X½.   *b*. Flower, X1¼.   *c*. Petal, X3.   *d*. Fruiting head, X2.   *e*. Achene, X5.

long to obovate, 8–12 mm long, sometimes more than and sometimes less than half as wide, usually more than twice as long as the sepals; stamens usually 40 or more; receptacle hispidulous; fruiting head globose to ovoid, 7–14 mm long, nearly as thick, with (5–) 10–20 achenes; achenes obovoid to nearly globose, flattish, 3.5–5.0 mm long, glabrous, with a high broad keel near the margin, the beak straight, 1.5–2.0 mm long.

COMMON NAME: Buttercup.

HABITAT: Low woods.

RANGE: Maryland to Nebraska, south to Texas and Florida.

ILLINOIS DISTRIBUTION: Champaign, Hancock, and Union counties.

This buttercup is similar to and readily confused with R. *septentrionalis*. Both species root at the nodes and have similar-appearing leaves and flowers.

The major distinction between these two species lies in the fruits. The achenes of R. *carolinianus* are very large (over 3.5 mm long) and have a conspicuous broad, high keel near the margin.

Benson (1948) reluctantly assigned specific status for R. *carolinianus* although seven years earlier he considered it to be R. *septentrionalis* var. *pterocarpus*. Fernald (1950) maintains R. *carolinianus* as a species and relegates R. *septentrionalis* var. *pterocarpus* to synonymy. On the other hand, Gleason (1952) retains both R. *carolinianus* and R. *septentrionalis* var. *pterocarpus* as distinct taxa.

Ranunculus carolinianus flowers during April and May.

22. Ranunculus repens L. Sp. Pl. 554. 1753.

Perennial stoloniferous herbs with fibrous roots; stems trailing or ascending, rooting at the nodes, branched, to 85 cm long, glabrous to hirsute; all leaves similar, deeply 3-parted or trifoliolate, the terminal segment stalked, glabrous or appressed-pubescent, borne on glabrous or appressed-pubescent petioles up to 25 cm long, often mottled with white; flowers several, up to 2 cm across, on appressed-pubescent or nearly glabrous pedicels up to 10 cm long; sepals 5, ovate, acute, spreading, greenish, pilose, 5–7 mm long; petals 5–numerous, bright yellow, obovate, 6–15 mm long, over half as wide, usually about twice as long as the sepals; stamens 5–many; receptacle pubescent; fruiting head subglobose, 6–10 mm in diameter, with 20–25 achenes; achenes obovoid, plump, 2.0–3.5 mm long, glabrous, narrowly margined, the beak recurved, 0.8–1.5 mm long.

Two varieties occur in Illinois.

1. Petals 5; stamens numerous _____ 22a. *R. repens* var. *repens*
1. Petals numerous; stamens few _____ 22b. *R. repens* var. *pleniflorus*

22a. Ranunculus repens L. var. **repens** *Fig. 37.*

Petals 5; stamens numerous.

COMMON NAME: Creeping Buttercup.
HABITAT: Roadsides, fields, and lawns.
RANGE: Native of the Old World; adventive from Labrador to Ontario, south to Missouri and North Carolina; western United States; Iceland; Costa Rica; Jamaica; Bermuda.
ILLINOIS DISTRIBUTION: Rare and apparently confined to the northern two-thirds of the state.

The creeping buttercup is similar to both *R. septentrionalis* and *R. carolinianus* in that the stems are often creeping and rooting at the nodes. *Ranunculus repens* differs from the other two by its plump achenes which have a beak never more than 1.5 mm long.

This native of the Old World is widely distributed as an adventive in North America, but it is not commonly found in Illinois.

Ranunculus repens flowers from May to August.

22b. Ranunculus repens L. var. **pleniflorus** Fern. Rhodora 19:138. 1917.

Petals numerous; stamens few.

COMMON NAME: Double-flowered Creeping Buttercup.
HABITAT: Waste ground.
RANGE: Probably originally from Europe; now planted as an ornamental and rarely escaped in some of North America.
ILLINOIS DISTRIBUTION: Jackson Co.: edge of campus of Southern Illinois University, Carbondale, August 14, 1967, *R. H. Mohlenbrock s.n.*

This is the garden form of the creeping buttercup in

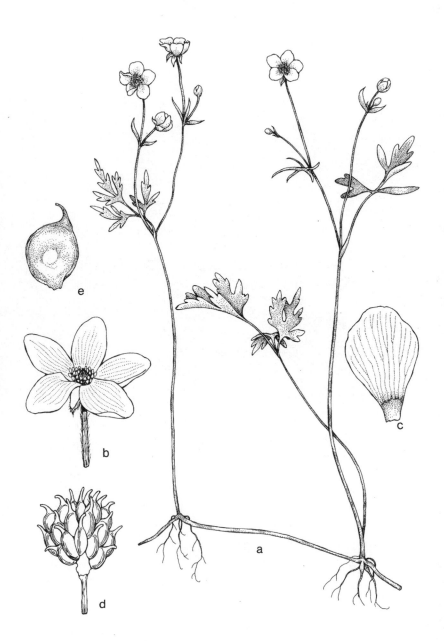

37. *Ranunculus repens* (Creeping Buttercup). *a*. Habit, X½. *b*. Flower, X2. *c*. Petal, X3. *d*. Fruiting head, X3. *e*. Achene, X6.

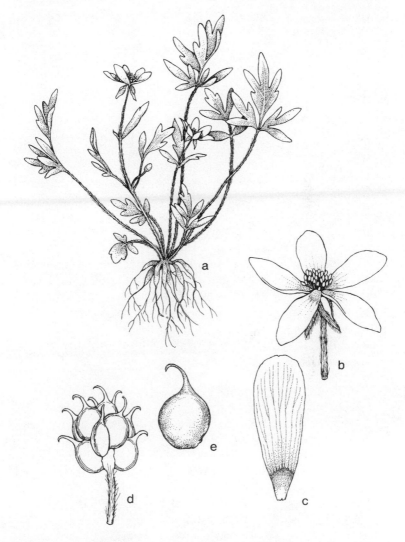

38. Ranunculus fascicularis (Early Buttercup). *a*. Habit, X½. *b*. Flower, X2½. *c*. Petal, X3¾. *d*. Fruiting head, X4. *e*. Achene, X5.

which most of the stamens have been transformed into petals, giving the "double-flowered" effect.

23. Ranunculus fascicularis Muhl. ex Bigel. Fl. Bost. ed. 1, 137. 1814. *Fig. 38.*

Ranunculus apricus Greene, Pittonia 4:145. 1900.

Ranunculus fascicularis var. *deforestii* Davis, Minn. Bot. Studies 2:470. 1900.
Ranunculus illinoensis Greene, Pittonia 5:195. 1903.
Ranunculus fascicularis var. *apricus* (Greene) Fern. Rhodora 38:178. 1936.

Perennial herbs from clusters of tuberous-thickened roots; stems erect, unbranched, often scapose, silky-canescent, to 30 cm tall; basal leaves shallowly or deeply lobed or 3– to 5–foliolate, rarely undivided, the divisions usually rather narrow, silky-canescent, on silky-canescent petioles up to 12 cm long; cauline leaves 1 or 2, smaller, 3– to 5–foliolate; flowers 1–4 per stem, up to 2 cm across, on silky-canescent pedicels up to 10 cm long; sepals 5, ovate, acute, spreading, greenish-yellow, densely silvery-canescent, 5–8 mm long; petals 5 (very rarely 9), bright yellow, elliptic to oblong, 7–15 mm long, less than half as wide, usually about twice as long as the sepals; stamens 40 or more; receptacle hispidulous; fruiting head subglobose, 5–8 mm in diameter, with 10–30 achenes; achenes obovoid to orbicular, flattened, 2.0–3.5 mm long, glabrous, narrow-margined, the beak straight or curved, 1.5–2.5 mm long.

COMMON NAME: Early Buttercup.
HABITAT: Open woods and meadows.
RANGE: New Hampshire to southern Ontario, south to Texas and Georgia.
ILLINOIS DISTRIBUTION: Throughout the state.

The early buttercup is distinguished by its fascicle of tuberous-thickened roots, its silky-canescent pubescence, and its generally dwarfed stature. The reduced number of leaves on the flowering stem is also characteristic.

Variation exists as to the degree of leaf-cutting. So-called typical specimens have some or all the basal leaves deeply parted. Other specimens, often referred to as var. *apricus* (Greene) Fern., have much less divided or even undivided leaves. Specimens such as this were collected by Baker on a hillside near Alto Pass, Union County, on April 21, 1900, and subsequently named *R. illinoensis* Greene. I am relegating both var. *apricus* (Greene) Fern. and *R. illinoensis* Greene to synonymy under *R. fascicularis* Muhl. since I can find no clear-cut demarcation of the characters in the specimens at hand.

An aberrant specimen with nine petals was collected on April

12, 1885, by Henry P. DeForest near Rossville, Vermilion County. Although it was subsequently described as *R. fascicularis* var. *deforestii* Davis, it is not worthy of distinction. In fact, some flowers on the type specimen have only five petals.

This buttercup flowers from early April through late May.

24. Ranunculus sardous Crantz. Stirp. Austr. ed. 1, fasc. 2:84. 1763. *Fig.* 39.

Ranunculus parvulus L. Mant. Pl. 79. 1767.

Annual with a sometimes swollen base and numerous fibrous roots; stems erect, much branched, hirsute to appressed-pilose, to 50 cm tall; basal leaves trifoliolate, the terminal segment stalked, pilose, on hirsute petioles up to 15 cm long; cauline leaves similar but usually with narrower divisions and with the uppermost becoming sessile; flowers numerous, to nearly 2 cm across, on slender strigose pedicels to 5 cm long; sepals 5, ovate, acute, reflexed, greenish-yellow, pilose, 2–5 mm long; petals 5, bright yellow, obovate, 8–12 mm long, nearly as broad, usually less than twice as long as the sepals; stamens 25 or more; receptacle pilose; fruiting head globose, to 8 mm in diameter, with 10 or more achenes; achenes nearly orbicular, flat, 2–3 mm long, the sides papillate or rarely smooth, distinctly narrow-margined, the beak curved, up to 0.5 mm long.

COMMON NAME: Buttercup.

HABITAT: Low fields and moist waste areas.

RANGE: Native of Europe; adventive at scattered localities in the United States and Canada.

ILLINOIS DISTRIBUTION: Confined to a few counties in the southernmost part of the state.

This species, *R. parviflorus*, and *R. arvensis* are the only Illinois representatives of section *Echinella*, characterized by the presence of papillae on the sides of the achenes.

In its growth form, *R. sardous* most nearly resembles *R. bulbosus*, even to the extent that some specimens of *R. sardous* are conspicuously swollen at the base.

A few specimens of *R. sardous* from Illinois are only sparsely papillate on the sides of the achenes.

The first collection in Illinois of this buttercup was made by W. M. Bailey and D. J. Hankla on October 2, 1948. Since that time, it has spread rapidly in wet fields of several southern Illinois counties. The showy flowers bloom in April and May.

39. *Ranunculus sardous* (Buttercup). *a*. Habit, X½. *b*. Flower, X2. *c*. Petal, X3. *d*. Fruiting head, X2½. *e*. Achene, X7½.

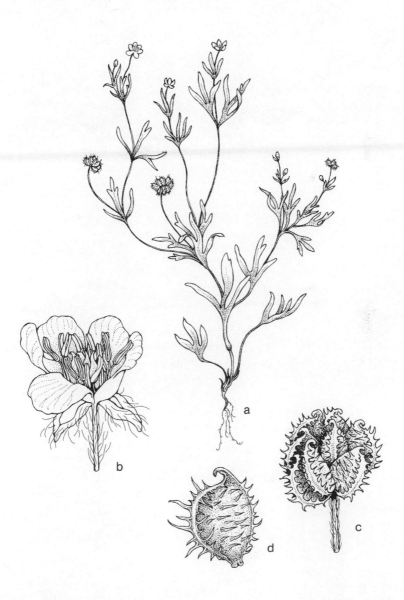

40. *Ranunculus arvensis* (Hunger-weed). *a*. Habit, X½. *b*. Flower, X3. *c*. Fruiting head, X3¾. *d*. Achene, X5.

25. Ranunculus arvensis L. Sp. Pl. 555. 1753. *Fig. 40.*

Annual from fibrous roots; stems erect, usually branched, glabrous or slightly pubescent, up to 30 cm tall (smaller in the Illinois material); some of the lowermost leaves often entire, broadly elliptic to oval, petiolate, glabrous or sparsely pubescent, other lowermost leaves deeply 3– to 7–cleft, petiolate; upper leaves 3– to 7–cleft, the lobes entire or toothed, sessile or nearly so; flowers solitary from the axils of the uppermost leaves, up to 1.6 cm across; sepals 5, free, green, acute, up to 5 mm long; petals 5, free yellow, obtuse, up to 8 mm long, longer than the sepals; stamens numerous; achenes 4 or more, flat, copiously spiny, up to 4.5 mm long, with a curved, sharply pointed beak up to 3 mm long.

COMMON NAME: Hunger-weed.

HABITAT: Along railroads.

RANGE: Native of Europe; rarely adventive in the United States.

ILLINOIS DISTRIBUTION: Jackson Co.: along Illinois Central Railroad, *D. Ladd & R. Tatina.*

This is a recent introduction in Illinois. Continued search along railroads should produce more records for this species.

The name Hunger-weed is derived from the fact that in Europe, this species grows abundantly in grain fields under conditions not conducive to the good growth of grains.

This species flowers from June to October.

2. *Caltha L.*–Marsh Marigold

Perennial herbs; stems floating, creeping, erect, or ascending, sometimes hollow; leaves basal and cauline, alternate, simple, entire or toothed; flowers yellow, white, or pink, usually solitary from the upper axils, on stout peduncles; sepals 5– 9, free, petallike, deciduous; petals absent; stamens numerous, free; pistils 3–15, free, with the several ovules in two rows; fruit a cluster of many-seeded follicles.

This is a genus of about fifteen wet ground or aquatic species, found primarily in the temperate and arctic regions of the world.

Several of the primitive characters of the family Ranunculaceae are exhibited well by this genus, such as free flower parts, numerous stamens and pistils, petallike sepals, and follicular fruits.

Only the following species occurs in Illinois.

41. *Caltha palustris* (Marsh Marigold).   *a*. Habit, in flower, X¾.   *b*. Flower, X1¼.
c. Fruit, X3.   *d*. Seed, X5.

1. **Caltha palustris** L. Sp. Pl. 558. 1753. *Fig. 41.*

Rather stout perennial from fibrous roots; stems ascending to erect, to 75 cm tall, glabrous, hollow, shallowly furrowed; basal leaves suborbicular to reniform, obtuse at the apex, cordate at the base, entire to dentate, glabrous, to 15 cm long and broad, the glabrous petioles longer than the blades; upper leaves similar but subcordate to truncate at the base, on very short petioles or sessile; flowers yellow, to 3.5 cm broad, solitary from the axils, on stout, glabrous peduncles to 10 cm long; sepals 5–9, yellow, oval, obtuse, to 2.5 cm long, to 2 cm broad; petals none; stamens numerous; pistils 3–12; follicles compressed, recurved-ascending, glabrous, to 1.5 cm long, prominently beaked, many-seeded.

COMMON NAME: Marsh Marigold; Cowslip.

HABITAT: Wet meadows; moist woods; springy calcareous fens.

RANGE: Labrador to Alaska, south to Nebraska and South Carolina; Europe; Asia.

ILLINOIS DISTRIBUTION: Confined to the northern half of the state, extending southward to Bond and Fayette counties.

The marsh marigold is readily distinguished by the simple, cordate leaves and the bright yellow, solitary flowers on stout axillary peduncles.

In its springy calcareous fen habitat of northeastern Illinois, it is associated with *Cardamine bulbosa, Carex hystricina, Eupatorium maculatum, E. perfoliatum, Lycopus americanus, Symplocarpus foetidus*, and others, according to Swink (1974).

Pioneers in bygone days used the young shoots of this plant, known to them as cowslip, for a spring vegetable.

The marsh marigold flowers from late March to June.

3. *Delphinium L.*–Larkspur

Annual or perennial herbs from fibrous or tuberous roots; stems erect, branched; leaves alternate, palmately divided; flowers showy, in terminal racemes or panicles; sepals 5, petallike, the upper one prolonged into a spur at the base; petals 2 or 4, smaller than the sepals, the upper two prolonged into spurs which protrude into the spurred sepal; stamens numerous; pistils 1–5, free, containing many ovules; fruit a cluster of 1–5 follicles, many-seeded.

There probably are at least 125 species in the genus, almost all confined to the north temperate regions of the world. In addition, many cultivated variants are known.

KEY TO THE SPECIES OF Delphinium IN ILLINOIS

1. Pistil 1; follicle 1; introduced annuals _____ 2
1. Pistils 3–5; follicles 3–5; native perennials _____ 3
   2. Pistil and follicle pubescent _____ 1. *D. ajacis*
   2. Pistil and follicle glabrous _____ 2. *D. consolida*
3. Stems glabrous; follicles spreading at maturity; seeds smooth _____
   _____ 3. *D. tricorne*
3. Stems pubescent; follicles erect at maturity; seeds roughened ____ 4
   4. Seeds winged, with appressed scales; some of the lower pedicels up
     to 2 cm long _____ 4. *D. carolinianum*
   4. Seeds unwinged, with projecting scales; none of the lower pedicels
     more than 1.5 cm long _____ 5. *D. virescens*

1. Delphinium ajacis L. Sp. Pl. 530. 1753. *Fig. 42.*

Annual from fibrous roots; stems slender, erect, simple or more often branched, to 1 m tall, puberulent; leaves deeply divided into linear or linear-filiform segments, the segments of the lower leaves somewhat wider than those of the upper, puberulent; flowers in terminal, loosely flowered racemes, blue, purple, pink, or white; at least some of the lower pedicels up to 20 mm long, puberulent; spurred sepal horizontal, slightly curved, to about 2 cm long; petals 2, united; pistil 1, pubescent; follicle 1, erect, short-beaked, pubescent, to 2 cm long; seeds marked with broken lines.

COMMON NAME: Rocket Larkspur.

HABITAT: Waste areas.

RANGE: Native of Europe; often grown as an ornamental in the United States.

ILLINOIS DISTRIBUTION: Occasional throughout the state.

    The rocket larkspur is a garden ornamental which sometimes is found as an escape in disturbed areas.

    This larkspur is unlike any other in Illinois, except *D. consolida*, because of its single pistil, single follicle, and two united petals. It differs from *D. consolida* by its pubescent pistil and follicle. The flowers, which are variously colored, bloom from June to August.

    This is probably the species that Mead, Vasey, Patterson,

42. *Delphinium ajacis* (Rocket Larkspur). *a*. Upper part of plant with flowers and fruits, X½. *b*. Flower, X1. *c*. Fruit, X2. *d*. Seed, X5.

Schneck, and Pepoon all reported as *Delphinium consolida* L. Higley and Raddin's report (1891) of *D. azureum* Michx. is probably an error for *D. ajacis.*

2. **Delphinium consolida** L. Sp. Pl. 530. 1753. *Fig. 43.*

Annual from fibrous roots; stems slender, erect, simple or more often branched, to 1 m tall, puberulent; leaves deeply divided into linear or linear-filiform segments, the segments of the lower leaves somewhat wider than those of the upper, puberulent; flowers in terminal, short, loosely flowered racemes, usually blue; none of the lower pedicels more than 15 mm long, puberulent; spurred sepal horizontal, slightly curved, to about 2 cm long; petals 2, united; pistil 1, glabrous; follicle 1, erect, short-beaked, glabrous, to 1.5 cm long; seeds marked with broken lines.

COMMON NAME: Larkspur.

HABITAT: Adventive on a landfill.

RANGE: Native of Europe; rarely introduced in the United States.

ILLINOIS DISTRIBUTION: DuPage Co.: on a landfill, August 3, 1973, *W. Crowley.*

This species, which is a common weed in Europe, is rarely found in the United States. Several early reports of this species from Illinois are actually based on *Delphinium ajacis.*

This larkspur blooms from May to August.

3. **Delphinium tricorne** Michx. Fl. Bor. Am. 1:314. 1803. *Fig. 44.*

Delphinium tricorne Michx. f. *albiflora* Millsp. Fl. W. Va. 322. 1892.

Perennial from a cluster of short, thick, tuberous roots; stems erect, simple, to 75 cm tall, glabrous; leaves mostly near the base of the plant, deeply 5– to 7–parted, the divisions again sometimes divided or toothed, glabrous or nearly so; flowers in terminal, loosely flowered racemes, purple (rarely white in f. *albiflora*), 2.5–3.5 cm long; pedicels arched-ascending, to 5 cm long, glabrous or hirtellous; spurred sepal nearly straight, except for a slight bend toward the tip, to 3 cm long; petals 4, free; pistils 3 (–5), glabrous; follicles 3 (–5), divergent, short-beaked, glabrous, reticulate, to 1.5 cm long; seeds smooth.

43. Delphinium consolida (Larkspur).   *a*. Upper part of plant, with flowers, X½.   *b*.
Flower, X1½.

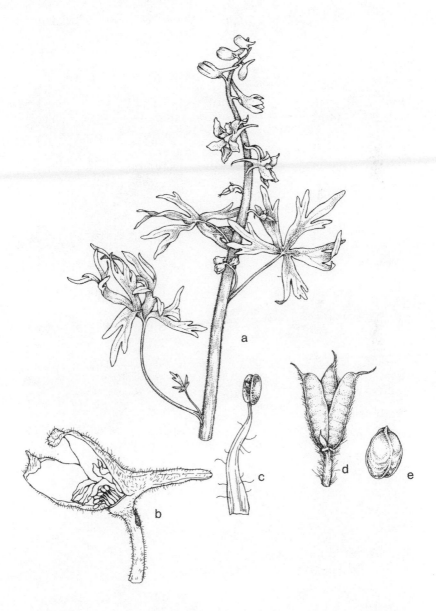

44. Delphinium tricorne (Dwarf Larkspur).   *a*. Upper part of plant, with flowers,
X½.   *b*. Flower, X1½.   *c*. Stamen, X10.   *d*. Fruit, X2.   *e*. Seed, X5.

COMMON NAME: Dwarf Larkspur.

HABITAT: Rich woods.

RANGE: Pennsylvania to Minnesota, south to Oklahoma and Georgia.

ILLINOIS DISTRIBUTION: Occasional to common in the southern three-fourths of Illinois; adventive in Cook County.

In rich woods of southern Illinois, this species may form massive colonies up to an acre in extent. During late April and early May, when the dwarf larkspur flowers, these woodlands are a solid mass of purple.

Pure white-flowered plants, sometimes designated as f. *albiflora* Millsp., have been found twice in the Pine Hills of Union County. A large colony of white-flowered plants has been seen near Sparta in Randolph County.

Delphinium tricorne differs from the somewhat similar *D. carolinianum* by its glabrous stems, its erect follicles, and its smooth seeds.

4. **Delphinium carolinianum** Walt. Fl. Car. 155. 1788. *Fig. 45.*

Delphinium azureum Michx. Fl. Bor. Am. 1:314. 1803.

Delphinium carolinianum Walt. var. *crispum* L. M. Perry, Rhodora 39:21. 1937.

Perennial from somewhat woody roots; stems erect, simple to sparingly branched, to 75 cm tall, puberulent to hirtellous, the pubescence along the upper part of the stem sometimes glandular; leaves basal and cauline, deeply 3– to 5–parted, the divisions again divided into linear segments, puberulent; flowers rather remote, in terminal racemes, deep purple, 2–3 cm long; pedicels ascending, to 3 cm long, pubescent; spurred sepal curved upward, to 1.8 cm long; petals 4, free, the lower 2–cleft; pistils mostly 3, puberulent; follicles 3, erect, slenderly beaked, puberulent, to 2 cm long; seeds winged, with appressed scales.

COMMON NAME: Wild Blue Larkspur.

HABITAT: Dry, often sandy soil.

RANGE: Virginia to southern Illinois and Oklahoma, south to Texas and Florida.

ILLINOIS DISTRIBUTION: Known only from Adams, Henderson, Macon, Mercer, Moultrie, and Pike counties.

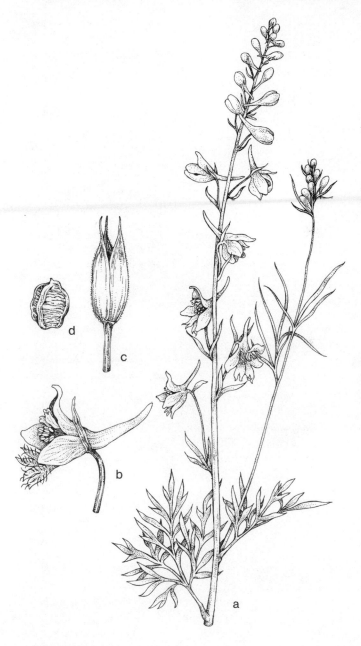

45. *Delphinium carolinianum* (Wild Blue Larkspur). *a*. Upper part of plant with flowers, X½. *b*. Flower, X1½. *c*. Fruit, X1½. *d*. Seed, X5.

The pubescence of the stems and leaves distinguishes this larkspur from *D. tricorne*. *Delphinium virescens*, which is also similar, has white flowers and unwinged seeds.

Those Illinois specimens lacking glandular pubescence have sometimes been designated var. *crispum* Perry.

Patterson first collected this species from Henderson County in 1873. This represented the only known record of this species from Illinois until 1944 when it was found in Moultrie County.

This wild blue larkspur blooms in May and June.

5. Delphinium virescens Nutt. Gen. 2:14. 1818. *Fig. 46.*

Delphinium albescens Rydb. Bull. Torrey Club 26:583. 1899.

Perennial from somewhat woody roots; stems erect, simple to sparingly branched, to 80 cm tall, glandular-pubescent in the upper part, pubescent in the lower part; leaves basal and cauline, deeply 3- to 5-parted, the divisions again divided into linear segments, pubescent; flowers rather densely flowered, in terminal racemes, white, 2–3 cm long; pedicels ascending, to 4 cm long, pubescent; spurred sepal straight or nearly so, to 1.8 cm long; petals 4, free; pistils mostly 3, puberulent; follicles 3, erect, slenderly beaked, puberulent, to 2 cm long; seeds unwinged, with projecting scales.

COMMON NAME: Prairie Larkspur.

HABITAT: Prairies.

RANGE: Manitoba to Colorado, south to Texas and western Illinois.

ILLINOIS DISTRIBUTION: Very rare; collected by S. B. Mead before 1850 in Hancock County.

This prairie species, probably extinct in Illinois, differs from the similar *D. carolinianum* by its whitish flowers and its more erect inflorescence.

The flowers are produced in May and June.

4. Trautvetteria Fisch. & Mey.–False Bugbane

Perennial herb; leaves basal and cauline, palmately lobed; inflorescence corymbose, with several small flowers; sepals 3–5, free, petallike, caducous; petals absent; stamens numerous, free; pistils numerous, free; ovaries superior, each with 1 ovule; achenes numerous, beaked.

Trautvetteria is composed of one species in the eastern United States, one in the western United States, and one in Asia. It has no close relatives among other Illinois members of the Ranunculaceae.

46. Delphinium virescens (Prairie Larkspur).   *a*. Lower part of stem with leaves,
X½.   *b*. Fruiting branch, X¼.   *c*. Flower, X2.   *d*. Seed, X12½.

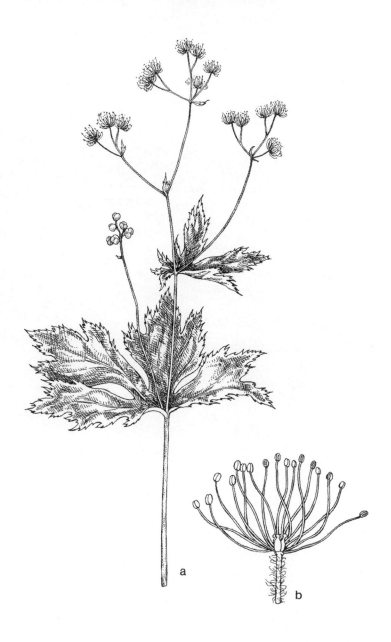

47. *Trautvetteria caroliniensis* (False Bugbane). *a*. Upper part of plant, X½. *b*. Flower, X4.

1. **Trautvetteria caroliniensis** (Walt.) Vail, Mem. Torrey Club
   2:42. 1890. *Fig. 47.*
Hydrastis caroliniensis Walt. Fl. Car. 156. 1788.
Cimicifuga palmata Michx. Fl. Bor. Am. 1:316. 1803.
Trautvetteria palmata (Michx.) Fisch. & Mey. Ind. Sem. Petr.
   1:22. 1834.
Trautvetteria palmata (Michx.) Fisch. & Mey. β *coriacea* Huth
   in Engl. Bot. Jahrb. 16:288. 1892.

Perennial herb from fibrous roots; stems stout, erect, glabrous, branched, to nearly 1 m tall; basal leaves palmately 5– to 11–lobed, to 35 cm long, usually a little broader, glabrous above, more or less puberulent beneath, the lobes sharply toothed, acute; petioles glabrous, to 50 cm long; cauline leaves alternate, similar to the basal leaves except for the progressively shorter petioles; flowers white, to 1.3 cm across; sepals (3–) 4 (–5), white, broadly ovate, concave, acute at the tip, glabrous, to 1 cm long, falling away as the flower opens; petals absent; stamens numerous, with slender filaments and oblong anthers; achenes numerous in small heads, beaked, more or less 4-angular, inflated, to 8 mm long.

COMMON NAME: False Bugbane.
HABITAT: Along rivers and streams.
RANGE: Pennsylvania to Missouri, south to Georgia and Florida.
ILLINOIS DISTRIBUTION: Known only from near Beardstown, Cass County, collected by *C. Geyer.*

The only known location of this species from Illinois was along a branch of the Sangamon River about three miles northereast of Beardstown, Cass County, in 1842. Although the specimens were attributed to S. B. Mead in Patterson's Catalogue, the collections were actually made by Carl A. Geyer. Geyer was a German botanist who worked in the Dresden (Germany) Botanic Garden before coming to the United States in 1835. He collected in the St. Louis area during the period 1840–42, before returning to Germany in 1845. Francis Eugene McDonald, amateur botanist in the Peoria area from 1883 to 1920, made efforts to rediscover the false bugbane at Geyer's original locality, but without success.

Geyer's collection was described as a new variety (*coriacea*) by Huth, but there seems to be little justification for this.

The false bugbane flowers in June and July.

5. *Thalictrum* L.–*Meadow Rue*

Perennial herbs, usually from creeping, scaly rootstocks; leaves basal and cauline, 2–3 ternately compound, on petioles dilated at the base; flowers in corymbs, racemes, or panicles, perfect, polygamous, or dioecious; sepals 4–5, free, petallike, caducous; petals absent; stamens numerous, free; pistils up to 15, free; ovaries superior, each with 1 ovule; achenes up to 15, beaked.

Thalictrum is a genus of nearly one hundred species, most of which are found in the north temperate regions of the world. About twenty-five species occur in the United States. Boivin (1957) has merged the genus *Anemonella* with *Thalictrum*.

KEY TO THE SPECIES OF Thalictrum IN ILLINOIS

1. Middle and upper leaves sessile _____ 2
1. Middle and upper leaves petiolate _____ 3. *T. dioicum*
   2. Lower surface of leaflets eglandular and glabrous _____ 4
   2. Lower surface of leaflets glandular or pubescent _____ 3
3. Lower surface of leaflets glandular _____ 1. *T. revolutum*
3. Lower surface of leaflets pubescent but eglandular 2. *T. dasycarpum*
   4. Margins of leaflets revolute, the lower surface of the leaflets strongly
     conspicuously reticulate _____ 1. *T. revolutum*
   4. Margins of leaflets flat, the lower surface of the leaflets weakly
     reticulate _____ 2. *T. dasycarpum*

1. **Thalictrum revolutum** DC. Syst. 1:173. 1818. *Fig.* 48.
Thalictrum revolutum DC. f. *glabrum* Pennell, Bartonia 12:12. 1931.
Thalictrum purpurascens DC. var. *ceriferum* Austin ex Gray, Man. Bot., ed. 5, 39. 1867.

Perennial herb from cordlike rootstocks; stems erect, to nearly 1 m tall, often purplish, glabrous or nearly so; leaves ternately decompound, the lower ones in glabrous petioles much longer than the upper petioles; leaflets firm, ovate to obovate, mostly 3-lobed near the apex, strongly reticulate-nerved, to 3 cm long, sometimes nearly as broad, dark green and more or less glabrous above, paler and glandular-pubescent or rarely glabrous beneath, the margins revolute; flowers white, polygamous or dioecious, the pistillate flowers usually with some stamens; sepals 4–5, white, ovate to oblong, glabrous, to 4 mm long; petals absent; stamens numerous, exserted, drooping shortly after the flowers open, the anthers linear; pistils glandular-pubescent; achenes up to 15 in a cluster,

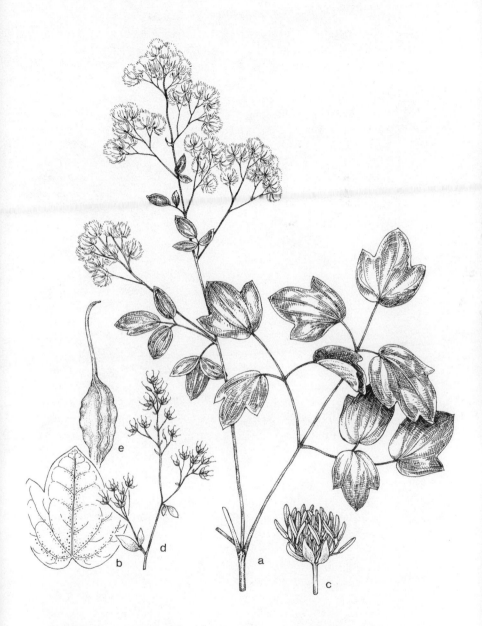

48. Thalictrum revolutum (Waxy Meadow Rue). *a.* Upper part of plant, with staminate flowers, X.675. *b.* Leaflet, X1⅛. *c.* Staminate flower, X2¼. *d.* Pistillate inflorescence, X.45. *3.* Fruit, X6¾.

narrowly ovoid, short-stipitate at the base, puberulent, ridged, to 6 mm long, including the slender, sometimes bent beak.

COMMON NAME: Waxy Meadow Rue.

HABITAT: Prairies; open woods; moist meadows.

RANGE: Massachusetts to southern Ontario, south to Missouri, Alabama, and northern Florida.

ILLINOIS DISTRIBUTION: Occasional throughout the state; possibly in every county.

The waxy meadow rue owes its waxy appearance to the glandular pubescence of the lower leaflet surface. These glands make the plant strongly aromatic. A few specimens have been found in Illinois in which these glands are lacking from the leaflets. These are sometimes designated f. *glabrum* Pennell.

The pistillate flowers usually have a few stamens present and also seem to possess slightly shorter petals than the other flowers.

The waxy meadow rue flowers from late May through June.

2. **Thalictrum dasycarpum** Fisch. & Lall. Ind. Sem. Hort. Petrop. 8:72. 1842.

Perennial herb from short, thick rootstocks; stems erect, to nearly 1 m tall, usually purplish, glabrous or pubescent; leaves ternately decompound, the lower ones on glabrous or puberulent petioles much longer than the upper petioles; leaflets firm or thin, oblong to obovate, mostly 3-lobed near the apex, weakly reticulate-nerved, to 4 cm long, sometimes nearly as broad, dark green and glabrous or sparsely puberulent above, somewhat paler and puberulent or glabrous beneath, the margins flat on more or less revolute; flowers whitish, mostly unisexual; sepals 4–5, lanceolate to narrowly obovate, glabrous, to 5 mm long; petals absent; stamens numerous, exserted, drooping shortly after the flowers open; pistils glabrous or pubescent; achenes up to 15 in a cluster, lanceoloid to ovoid, short-stipitate at the base, glabrous or pubescent, ridged, to 5 mm long, including the slender beak.

Two varieties, sometimes treated as species, occur in Illinois.

1. Leaflets firm, pubescent on the lower surface _____
   _____ 2a. *T. dasycarpum* var. *dasycarpum*
1. Leaflets thin, glabrous on the lower surface _____
   _____ 2b. *T. dasycarpum* var. *hypoglaucum*

49. *Thalictrum dasycarpum* (Purple Meadow Rue). *a*. Upper part of plant with staminate flowers, X½. *b*. Staminate flower, X3½. *c*. Pistillate inflorescence, X½. *d*. Fruit, X7½.

50. *Thalictrum dasycarpum* var. *hypoglaucum* (Meadow Rue). *a*. Upper part of plant, with staminate flowers, X¾. *b*. Staminate flower, X3½. *c*. Pistillate inflorescence, X½. *d*. Fruit, X7½.

2a. **Thalictrum dasycarpum** Fisch. & Lall. var. **dasycarpum**
Fig. 49.

Leaflets firm, pubescent on the lower surface.

COMMON NAME: Purple Meadow Rue.

HABITAT: Moist meadows; moist wooded ravines.

RANGE: Maine to Alberta, south to Arizona, Kansas, and southern Indiana.

ILLINOIS DISTRIBUTION: Occasional throughout the state.

The firm, pubescent leaflets provide easy ways to distinguish this variety from var. *hypoglaucum.* In addition to these leaflet characters, Fernald (1950) reports overlapping characteristics relating to filament length, anther length, and stigma length, but these are not significant enough to justify different species.

The early Illinois botanists referred our species to *T. purpurascens* L., but this latter binomial applies to a different species.

The purple meadow rue blooms from May to early June.

2b. **Thalictrum dasycarpum** Fisch. & Lall. var. **hypoglaucum** (Rydb.) Boivin, Rhodora 46:482. 1944. *Fig. 50.*

Thalictrum hypoglaucum Rydb. Brittonia 1:88. 1931.

Leaflets thin, glabrous on the lower surface.

COMMON NAME: Meadow Rue.

HABITAT: Moist meadows; moist wooded ravines.

RANGE: Minnesota to South Dakota, south to Arizona and Louisiana; reported from British Columbia (Fernald, 1950).

ILLINOIS DISTRIBUTION: Occasionally scattered throughout the state.

Before Rydberg named this plant in 1931, it had been referred to by Illinois botanists either as *Thalictrum cornuti* or *T. polygamum*, but both these binomials apply to other species.

The meadow rue flowers from the last of May into July.

3. **Thalictrum dioicum** L. Sp. Pl. 545. 1753. *Fig. 51.*

Perennial dioecious herb from short, thick rootstocks; stems erect, to nearly 1 m tall, green or occasionally glaucous, glabrous; leaves

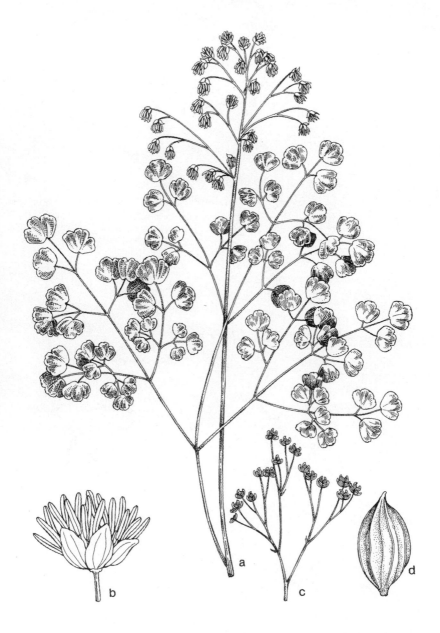

51. *Thalictrum dioicum* (Early Meadow Rue). *a*. Upper part of plant, with staminate flowers, X⅙. *b*. Staminate flower, X2½. *c*. Pistillate inflorescence, X½. *d*. Fruit, X6.

ternately decompound, with both the basal and the cauline leaves petiolate; leaflets thin, reniform to obovate to suborbicular, 5– to 7–lobed, not conspicuously reticulate-nerved, to 4 cm long, nearly as broad or slightly broader, glabrous, green above, usually paler and sometimes glaucous beneath, the margins usually not revolute; flowers greenish, unisexual, in terminal or axillary, spreading or pendent panicles; sepals 4–5, oblong to oval, glabrous, to 4.5 mm long; petals absent; stamens several, pendulous, with linear anthers; pistils glabrous; achenes up to 15 in a cluster, broadly ellipsoid, sessile or nearly so, glabrous, strongly ribbed, beakless, to 4.5 mm long.

COMMON NAME: Early Meadow Rue.

HABITAT: Moist woods; wooded clay slopes.

RANGE: Labrador to Saskatchewan, south to Missouri and Alabama.

ILLINOIS DISTRIBUTION: Occasional throughout the state.

When this species flowers during April and most of May, the leaves are not fully expanded. It has usually finished flowering by the time other members of the genus come into flower.

In addition to its usual habitat in moist woodlands, this species occurs on wooded clay slopes (Swink, 1974).

6. Actaea L.–Baneberry

Perennial herbs; stems erect; leaves 2–3 ternately compound; inflorescence terminal, racemose; sepals 3–5, free, petallike, caducous; petals 4–10, free, small; stamens numerous, free; pistil 1, the ovary superior, with many ovules; fruit a berry, with 2 rows of 3–several seeds.

Actaea is a genus of six species, all occurring in the north temperate regions of the world.

KEY TO THE SPECIES OF Actaea IN ILLINOIS

1. Pedicels much narrower than the peduncles in fruit; seeds 10 or more per berry, up to 4 mm long _____ 1. *A. rubra*
1. Pedicels about as thick as the peduncles in fruit; seeds 3–9 (–10) per berry, over 4 mm long _____ 2. *A. pachypoda*

1. Actaea rubra (Ait.) Willd. Enum. 561. 1809. *Fig. 52.*

Actaea spicata L. var. *rubra* Ait. Hort. Kew. 2:221. 1789.

52. *Actaea rubra* (Red Baneberry). *a*. Upper part of plant, with flowers, X1. *b*. Flower, X2½. *c*. Cluster of fruits, X½.

Perennial herb from slender roots; stems erect, usually branched, to 65 cm tall, glabrous or pubescent; leaves ternately decompound, the lower on petioles longer than the petioles of the upper; leaflets ovate to obovate, shallowly or deeply toothed, glabrous or pubescent, petiolulate; racemes ovoid, several-flowered; flowers white, on filiform, pubescent pedicels to 15 mm long; sepals 3–5, whitish, falling away as the flower opens; petals 4–10, white, spatulate, shorter than the stamens; stamens numerous, with slender white filaments; pistil glabrous; berries red or white, oval or ellipsoid, shiny, to 1.3 cm long, on pedicels much narrower than the peduncles, with 10 or more seeds 3–4 mm long.

Two forms are known from Illinois.

1. Berries red _____ 1a. A. rubra f. rubra
1. Berries white _____ 1b. A. rubra f. neglecta

1a. Actaea rubra (Ait.) Willd. f. rubra

Berries red.

COMMON NAME: Red Baneberry.
HABITAT: Moist woodlands; springy calcareous areas.
RANGE: Labrador to Alaska, south to Oregon, Colorado, northern Illinois, and New Jersey.
ILLINOIS DISTRIBUTION: Restricted to the northern one-fourth of the state.

Most specimens of Actaea rubra in Illinois have red berries, rather than white. The berries are somewhat poisonous, and are produced from August to October.

The flowers appear during late April and May.

1b. Actaea rubra (Ait.) Willd. f. neglecta (Gillman) Robins. Rhodora 10:66. 1908.

Actaea neglecta Gillman ex Lloyd, Drugs & Medicines 235. 1884–85.

Berries white.

COMMON NAME: White-fruited Red Baneberry.
HABITAT: Moist woodlands.
RANGE: Same as f. rubra.
ILLINOIS DISTRIBUTION: Rare in the northern one-fourth of the state, absent elsewhere.

The white-fruited form of the red baneberry is less common than the red-fruited form in Illinois. It resembles A. pachypoda, but differs by its more slender pedicels and its smaller and fewer seeds.

53. *Actaea pachypoda* (Doll's-eyes). *a*. Upper part of plant, with fruits, X½. *b*. Inflorescence, X½. *c*. Flower, X1¼. *d*. Fruit, X2. *e*. Seed, X5.

2. **Actaea pachypoda** Ell. Bot. S.C. & Ga. 2:15. 1821. *Fig.* 53.
Actaea alba Bigel. In Eaton, Man. ed. 2:123. 1818, non Mill. (1768).

Perennial herb from slender roots; stems erect, usually branched, to 65 cm tall, glabrous or pubescent; leaves ternately decompound, the lower on petioles longer than the petioles of the upper; leaflets ovate to obovate, shallowly or deeply sharply toothed, glabrous or pubescent, petiolulate; racemes ellipsoid, several-flowered; flowers white, on rather stout, nearly glabrous pedicels to 15 mm long; sepals 3–5, whitish, falling away as the flower opens; petals 4–10, white, spatulate, shorter than the stamens; stamens numerous, with slender white filaments; pistil glabrous; berries white, with a purple tip, globose to ovoid, shiny, about 1 cm in diameter, on pedicels about as thick as the peduncles, with 3–9 (–10) seeds 4–5 mm long.

COMMON NAME: Doll's-eyes.
HABITAT: Moist woods.
RANGE: Nova Scotia to Manitoba, south to Oklahoma and Georgia.
ILLINOIS DISTRIBUTION: Throughout the state.

The white color of the flowers of this and the preceding species is due for the most part to the white filaments.

Doll's-eyes is an appropriate name for this plant because the shiny white fruits tipped by the flat, purple stigma certainly resemble eyes. The berries are mildly poisonous. The rhizomes have reputed medicinal value.

The flowers appear in April and May, while the fruits mature from July to October.

Fernald (1940) has given considerable evidence that the correct name for this species is *A. pachypoda*, although most botanists for several decades have used the name *A. alba*. Elliott's use of *A. alba* for this species is pre-dated by Miller's use of the same binomial for a different taxon. Thus *A. pachypoda* appears to be correct.

7. Cimicifuga L.–Bugbane

Perennial herbs; stems erect; leaves 2–3 ternately compound; inflorescence terminal and axillary, racemose; sepals 2–5, free, petal-like, caducous; petals 0–9, free, bilobed at the tip; stamens numerous, free; pistils 1–8, the ovaries superior, with several ovules; fruit a cluster of 1–8 follicles.

Cimicifuga is composed of about a dozen species native to North America, Asia, and eastern Europe. It differs from the very similar *Actaea* by its several inflorescences and follicular fruits.

A revision of the genus has been made by Ramsey (1964).

KEY TO THE SPECIES OF Cimicifuga IN ILLINOIS

1. Pistil 1 (–2), sessile or nearly so; seeds not scaly _____ 2
1. Pistils 3–8, stipitate; seeds scaly _____ 3. *C. americana*
    2. At least the terminal leaflet cordate _____ 1. *C. rubifolia*
    2. All leaflets truncate, subcordate, or cuneate ____ 2. *C. racemosa*

    1. **Cimicifuga rubifolia** Kearney, Bull. Torrey Club 24:561. 1897. *Fig. 54.*
Cimicifuga cordifolia Pursh, Fl. Am. Sept. 373. 1814.
Cimicifuga racemosa (L.) Nutt. var. *cordifolia* (Pursh) Gray, Syn. Fl. 1(1):55. 1895.

Perennial herb from thick, knotty rhizomes; stems erect, to 2 m tall, glabrous; leaves ternately decompound; leaflets up to 9 in number, broadly ovate, obtuse to acute to acuminate at the apex, cordate at the base, coarsely toothed, to 25 cm long, often nearly as broad, glabrous; racemes terminal and lateral, erect, the terminal up to 75 cm long; flowers whitish, to 1.5 cm broad, on slender pedicels; sepals 4–5, whitish, soon falling; petals 4–8, usually bilobed at the apex, usually shorter than the stamens; stamens numerous, with whitish filaments; pistil 1, glabrous, sessile; follicle broadly oval, beaked, to 1 cm long, glabrous, with several smooth, flattened seeds.

COMMON NAME: Heart-leaved Black Cohosh.
HABITAT: Rich woodlands.
RANGE: North Carolina and Tennessee to Georgia; southern Illinois.
ILLINOIS DISTRIBUTION: Restricted to Gallatin, Hardin, Jackson, Johnson, Massac, and Pope counties.
    Until Ramsey's (1964) study of the genus *Cimicifuga*, this species often was considered only a variety of *C. racemosa*.

54. Cimicifuga rubifolia (Heart-leaved Black Cohosh). *a*. Leaves, X.45. *b*. Flower, X2.7. *c*. Cluster of fruits, X.45. *d*. Fruit, X2¼. *e*. Seed, X4½.

The best field character for *C. rubifolia* is the cordate terminal leaflet.

This species flowers in July and August, usually slightly later than *C. racemosa*.

The first Illinois collection was made on August 17, 1951, by Julius R. Swayne from Jackson Hollow, Pope County. The Illinois stations are a considerable distance away from the nearest localities in Tennessee.

2. Cimicifuga racemosa (L.) Nutt. Gen. 2:15. 1818. *Fig. 55.*

Actaea racemosa L. Sp. Pl. 504. 1753.

Perennial herb from thick, knotty rhizomes; stems erect, to 2 m tall, glabrous, branched; leaves ternately decompound, sometimes ultimately 5-parted; leaflets usually more than 9 in number, ovate to oblong to narrowly obovate, obtuse to acute at the apex, subcordate, truncate, or cuneate at the base, coarsely toothed, to 10 cm long, to 7 cm broad, glabrous; racemes terminal and lateral, erect, the terminal up to 75 cm long; flowers whitish, to 1.5 cm broad, on slender pedicels; sepals 4–5, whitish, soon falling; petals 4–8, usually bilobed at the apex, usually shorter than the stamens; stamens numerous, with whitish filaments; pistil 1 (–2), glabrous, sessile; follicle broadly oval, minutely beaked, to 1 cm long, glabrous, with several smooth, flattened seeds.

COMMON NAME: Black Cohosh.

HABITAT: Woods.

RANGE: Massachusetts to southern Ontario, south to Missouri and Georgia.

ILLINOIS DISTRIBUTION: Known only from Carroll, Kendall, Lake, St. Clair, and Wabash counties.

The native localities in St. Clair and Wabash counties are probably no longer in existence since the plant has not been collected there since late in the nineteenth century. The Carroll and Lake county records are based on introductions.

It flowers slightly earlier than *C. rubifolia* Kearney, blooming in June and July.

3. Cimicifuga americana Michx. Fl. Am. 1:316. 1803. *Fig. 56.*

Perennial herbs from thick, knotty rhizomes; stems erect, to 2 m tall, glabrous; leaves ternately decompound; leaflets ovate to ovate-

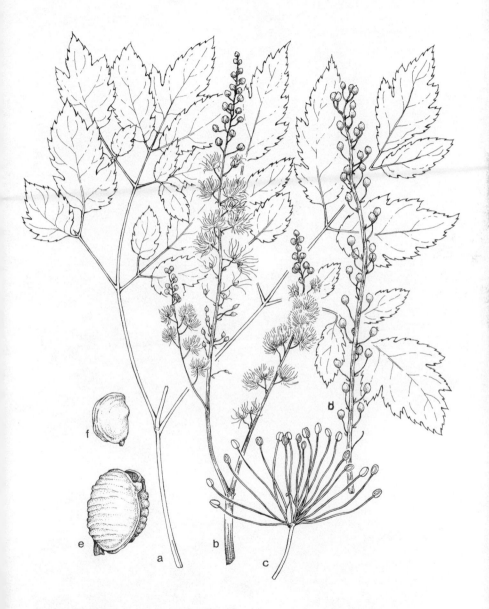

55. *Cimicifuga racemosa* (Black Cohosh). *a*. Leaves, X.45. *b*. Inflorescence, X.45. *c*. Flower, X3.15. *d*. Cluster of fruits, X.02½. *e*. Fruit, X2¼. *f*. Seed, X4½.

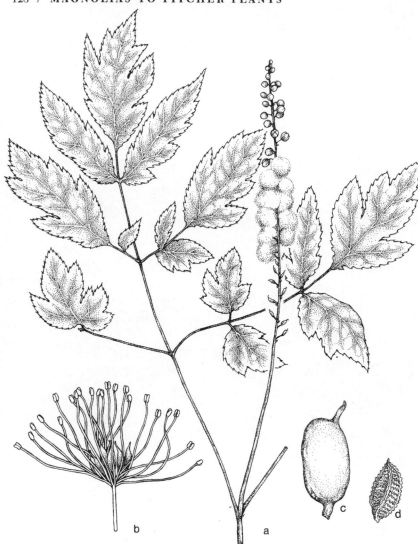

56. *Cimicifuga americana* (American Bugbane). *a.* Inflorescence and leaves, X½. *b.* Flower, X2½. *c.* Fruit, X2½. *d.* Seed, X5.

oblong, acute to acuminate at the apex, coarsely toothed, to 4.5 cm long, glabrous, the terminal leaflet usually cuneate at the base; racemes terminal and lateral, slender, the rachis puberulent, the terminal up to 60 cm long; flowers whitish, to 1.2 cm across, on slender pedicels; sepals 4–5, whitish, soon falling; petals 4–8, usually bilobed at the apex, with a basal nectary; stamens numerous, with whitish filaments; pistils 3–8, glabrous, stipitate; follicles 3–8, in a cluster, stipitate, flattened, to 1 cm long, glabrous, the beak subulate, the seeds densely scaly.

COMMON NAME: American Bugbane.

HABITAT: North-facing mesic talus slope.

RANGE: New York to Pennsylvania, south to Tennessee and Georgia; Illinois.

ILLINOIS DISTRIBUTION: Carroll Co.: Mississippi Palisades State Park, July 1, 1976 (leaf), September 20, 1976 (fruit), *Marlin Bowles.*

The remarkable discovery of this eastern mountainous species in northwestern Illinois raises much conjecture as to its status in Illinois. The several follicles in a cluster and the scaly seeds leave no doubt that the plant is *Cimicifuga americana.*

The Illinois locality is on a north-facing mesic talus slope (limestone), associated with *Trillium erectum, T. grandiflorum, Lycopodium lucidulum,* and *Osmunda claytonii. Quercus rubra* dominates the canopy. There is no evidence that the area has been associated with an old homestead.

Dr. Gwynn Ramsey, in personal communication, informs me that no native sites of *C. americana* are known from such a low elevation as that at Mississippi Palisades State Park.

This species flowers during August and September.

8. *Hepatica Mill.*–Hepatica

Perennial herbs; leaves all basal, evergreen, 3–lobed; flower solitary on a slender scape, subtended by three sessile bracts resembling sepals; sepals 5–12, free, petallike; petals absent; stamens numerous, free; pistils 2–12, free, the ovaries superior; achenes beaked.

Three species indigenous to the north temperate regions of the world comprise the genus. Steyermark and Steyermark (1960) have analyzed the North American taxa. As a result of their study, only the following variable species occurs in Illinois.

1. **Hepatica nobilis** Schreb. Spicil. Fl. Lips. 39. 1771.

Anemone hepatica L. Sp. Pl. 538. 1753.

Hepatica triloba Gilib. Fl. Lith. 2:273. 1781.

Perennial herbs from fibrous roots; leaves basal, triangular-reni-form, deeply cordate, 3–lobed, the lobes obtuse to acute, pubescent, to 6 cm long, often a little broader, on villous petioles to 20 cm long; flowering stalk villous, to 20 cm long; flower solitary, to 2.4 cm broad, subtended by 3 bracts, the bracts green, lanceolate to elliptic to oval, obtuse to acute, pubescent, to 18 mm long, to 10 mm broad; sepals 5–12, petallike, blue, purple, or white, oval to oblong, obtuse, pubescent, to 18 mm long, to 10.5 mm broad; petals absent; stamens numerous, shorter than the sepals; pistils 2–12, pubescent; achenes up to 12 in a cluster, fusiform to conic-ovoid, pubescent, slenderly beaked, to 5 mm long.

Two varieties occur in Illinois, both differing to some extent from typical var. *nobilis* of Europe.

KEY TO THE VARIETIES OF Hepatica nobilis IN ILLINOIS

1. Lobes of leaf obtuse; bracts obtuse _____ 1a. *H. nobilis* var. *obtusa*
1. Lobes of leaf acute; bracts acute _____ 1b. *H. nobilis* var. *acuta*

1a. Hepatica nobilis Schreb. var. **obtusa** (Pursh) Steyermark in Steyermark & Steyermark, Rhodora 62:232. 1960. *Fig. 57.*

Hepatica triloba Gilib. var. *obtusa* Pursh, Fl. Am. Sept. 2:391. 1814.

Hepatica triloba Gilib. var. *americana* DC. Syst. Nat. 1:216. 1817.

Hepatica americana (DC.) Ker in Edwards, Bot. Reg. 5:t. 387. 1819.

Lobes of leaf obtuse; bracts obtuse.

COMMON NAME: Round-lobed Liverleaf.

HABITAT: Rich woods, mostly on acidic soils; dune slopes.

RANGE: Nova Scotia to Manitoba, south to Missouri and Florida.

ILLINOIS DISTRIBUTION: Restricted to the extreme northeastern counties.

Although most botanists have considered this taxon to be a distinct species, Steyermark and Steyermark (1960) give convincing arguments why it should be considered a variant of the European *H. nobilis*.

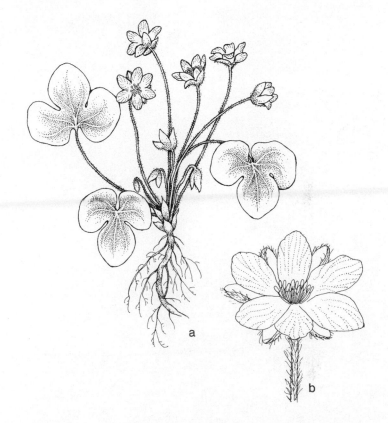

57. *Hepatica nobilis* var. *obtusa* (Round-lobed Liverleaf). *a*. Habit, X½. *b*. Flower, X1½.

The several variations in sepal color have given rise to numerous named lesser taxa, none of which is considered significant here.

The flowers appear in April and early May.

1b. Hepatica nobilis Schreb. var. **acuta** (Pursh) Steyermark in Steyermark & Steyermark, Rhodora 62:232. 1960. *Fig. 58*.

Hepatica triloba Gilib. var. *acuta* Pursh Fl. Am. Sept. 2:391. 1814.

Hepatica triloba Gilib. var. *acutiloba* (DC.) Warne, Am. Ent. & Bot. 2:313. 1870.

Anemone acutiloba (DC.) Laws. Rev. Can. Ranunc. 30. 1870.

Lobes of leaf acute; bracts acute.

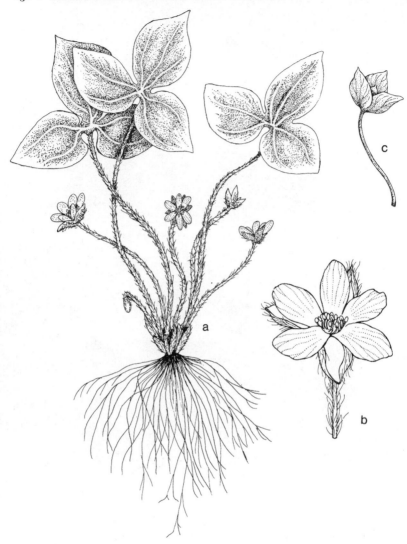

58. *Hepatica nobilis* var. *acuta* (Sharp-lobed Liverleaf).   *a*. Habit, X½.   *b*. Flower, X1½.   *c*. Fruit, X1½.

COMMON NAME: Sharp-lobed Liverleaf.

HABITAT: Rich woods.

RANGE: Quebec to Minnesota, south to Missouri and Georgia.

ILLINOIS DISTRIBUTION: Occasional throughout the state.

In the few areas in Illinois where var. *obtusa* and var. *acuta* occur together, some specimens exist which show definite intergradation. In such areas, Steyermark and Steyermark have observed that var. *obtusa* occurs on more acid, leached soils on top of ravines and slopes, while var. *acuta* occurs in the richer soils of the creek bottom.

This variety blooms from mid-March to early May.

9. *Hydrastis Ellis*–Goldenseal

Perennial herbs; leaves palmately lobed; flower solitary; sepals 3, free, petallike, caducous; petals absent; stamens numerous, free; pistils numerous, free, the ovaries superior, each with 2 ovules; fruit a head of berries.

In addition to the North American species described below, *Hydrastis* has one other species in Japan.

1. Hydrastis canadensis L. Syst., ed. 10, 1088. 1759. *Fig.* 59.

Perennial herb from a knotty yellow rhizome; basal leaf one, palmately 5– to 9–lobed, to 20 cm long at maturity, nearly as broad, pubescent, on pubescent petioles to 20 cm long; cauline leaves 2, similar in shape to the basal leaf but progressively smaller and on shorter petioles; flower solitary from the axil of the uppermost leaf, to 1.2 cm broad, on a pubescent peduncle to 3 cm long; sepals 3, whitish, lanceolate, about 1 mm long, falling away as the flower opens; petals none; stamens numerous, the filaments whitish, widened to about 5 mm long; pistils 12–20, glabrous or nearly so; fruiting head ovoid, to nearly 2 cm long, composed of 12–20 densely clustered berries, each berry red, beaked, 1– to 2–seeded.

59. *Hydrastis canadensis* (Goldenseal). *a*. Habit, in flower, X½. *b*. Flower, X3½.
c. Leaf, with fruits, X½. *d*. Seed, X5.

COMMON NAME: Goldenseal.

HABITAT: Rich woods.

RANGE: Vermont to Minnesota, south to Nebraska and Georgia.

ILLINOIS DISTRIBUTION: Occasional throughout the state.

The yellow knotty rhizomes of the goldenseal contain substances which were important for medicinal purposes in the past. As a result, the digging of these rhizomes has caused the plant to be far less common than it used to be in Illinois.

The fruiting head of red berries, which resembles a blackberry, is attractive with the dark green upper leaf as a background.

The goldenseal flowers in April and May.

10. *Isopyrum* L.–*False Rue Anemone*

Perennial herbs from sometimes tuberlike roots; leaves 2–3, ternately compound; flowers solitary or in panicles; sepals 5, free, petallike; petals 5, free, or absent; stamens numerous, free; pistils 2–20, free, the ovaries superior, each with several ovules; fruit a head of follicles.

Isopyrum is a genus of fifteen species, all occurring in north temperate regions. It is similar in general appearance to *Anemonella*, but differs by its head of fruiting follicles and its basal and alternate leaves. *Anemonella*, on the other hand, has an umbel of achenes, basal leaves, and a whorl of bracteal leaves.

Only the following species occurs in Illinois.

1. **Isopyrum biternatum** (Raf.) Torr. & Gray, Fl. N. Am. 1:660. 1840. *Fig. 60.*

Enemion biternatum Raf. Journ. Phys. 91:70. 1820.

Herbaceous perennial from fibrous and sometimes tuberlike roots; stems slender, erect, glabrous, to 30 cm tall; basal leaves 2–3 ternately divided, long-petiolate; leaflets thin, obovate, 3–lobed, to 2.5 cm long, about as broad, glabrous, the lobes obtuse; upper leaves alternate, with short petioles, the leaflets similar in shape to the leaflets of the basal leaves; flowers 1–several, terminal and from the upper axils, to 2 cm broad; sepals 5, white, oblong to obovate, obtuse, to 15 mm long; petals none; stamens numerous, with white filaments; pistils 2–10, glabrous; follicles 3–6, ovoid, beaked, glabrous, to 5 mm long, with 2 or more seeds.

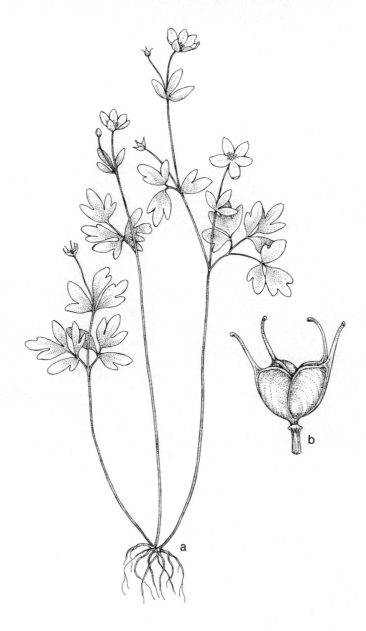

60. *Isopyrum biternatum* (False Rue Anemone). *a*. Habit, X½. *b*. Fruits, X5.

COMMON NAME: False Rue Anemone.

HABITAT: Moist woods.

RANGE: Southern Ontario to Minnesota, south to Texas and Florida.

ILLINOIS DISTRIBUTION: Common throughout the state.

This delicate species is common in rich woods in all parts of the state. It flowers from March to May.

11. *Anemonella Spach.*–Rue Anemone

Perennial herb with tuberlike roots; leaves all basal, 2–3 ternately compound; flowers in terminal umbels borne from a whorl of involucral leaves; sepals 5–10, free, petallike; petals absent; stamens numerous, free; pistils 4–15, free, the ovaries superior, each with a single ovule; fruit a cluster of achenes.

Although this monotypic genus resembles *Isopyrum*, it is more closely related to *Thalictrum*. In fact, Boivin (1957) proposes that the two genera be merged. While there is some merit to this union, I do not choose to follow it at this time.

Only the following species comprises the genus.

1. **Anemonella thalictroides** (L.) Spach, Hist. Veg. 7:240. 1839. *Fig. 61*.

Anemone thalictroides L. Sp. Pl. 542. 1753.

Thalictrum anemonoides Michx. Fl. Bor. Am. 1:322. 1803.

Syndesmon thalictroides (L.) Hoffmg. Flora 15(2):Intell. Bl. 4, 34. 1832.

Anemonella thalictroides (L.) Spach f. *favilliana* Bergseng ex Fassett, Trans. Wisc. Acad. Sci. 38:199. 1946.

Thalictrum thalictroides (L.) Boivin, Bull. Soc. Roy. Bot. Belg. 89:3 1957.

Herbaceous perennial from a cluster of tuberlike roots; leaves all basal, 2–3 ternately compound, on petioles to 25 cm long; leaflets thin, obovate to oblong to nearly orbicular, shallowly 3–lobed, to 4 cm long, nearly as broad, glabrous, the lobes obtuse; flowering stem glabrous, erect, slender, to 35 cm long, with a whorl of involucral leaves at the tip, the leaves simple or ternately divided, the leaflets similar to those of the basal ones; flowers few in an umbel, to 2 cm across, on slender, glabrous pedicels; sepals 5 or 10, white or pinkish, oval, obtuse, to 15 mm long; petals none; stamens numerous, with white filaments, rarely petallike (in f. *favilliana*); pis-

61. *Anemonella thalictroides* (Rue Anemone). *a*. Habit, X½. *b*. Flower, X2½. *c*. Fruiting head, X1½.

tils 4–15, glabrous; achenes 4–15, ovoid, short-beaked, glabrous, to 12 mm long, 1-seeded.

COMMON NAME: Rue Anemone.

HABITAT: Dry wooded slopes; moist woods.

RANGE: Maine to Minnesota, south to Oklahoma and Florida.

ILLINOIS DISTRIBUTION: Occasional throughout the state.

This species is similar to the false rue anemone, differing by the whorl of involucral leaves and the umbel of achenes.

A form with the stamens transformed into petals has been collected in Lake County. It has been designated as f. *favilliana* Bergseng.

The rue anemone flowers during April, May, and June.

12. Anemone L.–*Anemone*

Perennial herbs with slender or tuberlike roots or rhizomes; leaves all basal, palmately lobed or divided; flowers solitary or in umbels, borne from a whorl of involucral leaves; sepals 4–numerous, free, petallike; petals absent; stamens numerous, free; pistils 2–several, free, the ovaries superior, each with a single ovule; fruit a cluster of achenes.

Anemone is a genus of nearly one hundred species in the temperate and subarctic regions of the Northern and Southern hemispheres.

KEY TO THE SPECIES OF Anemone IN ILLINOIS

1. Styles 2–4 cm long, plumose; staminodia present _____ 1. *A. patens*
1. Styles up to 4 mm long, usually pubescent but not plumose; staminodia absent _____ 2
   2. Plants arising from a tuber; sepals 10–20 _____ 2. *A. caroliniana*
   2. Plants arising from rhizomes; sepals 5 (–6) _____ 3
3. Leaves of the involucre sessile; beak of achene 2–5 mm long _____ _____ 3. *A. canadensis*
3. Leaves of the involucre petiolate; beak of achene less than 2 mm long _____ 4
   4. Basal leaf solitary; plants at maturity less than 30 cm tall; achenes hirsutulous, but not woolly _____ 4. *A. quinquefolia*
   4. Basal leaves 2–several; plants at maturity more than 30 cm tall; ach-

enes woolly -- 5
5. Leaves of the involucre 5–9; fruiting heads more than twice as long as wide; styles less than 1 mm long -------------- 5. *A. cylindrica*
5. Leaves of the involucre 3; fruiting heads less than twice as long as wide; styles 1 mm long or longer ------------------- 6. *A. virginiana*

  1. **Anemone patens** L. Sp. Pl. 538. 1753. *Fig. 62.*

  Pulsatilla patens (L.) Mill. Gard. Dict. ed. 8, no. 4. 1768.

  Anemonella ludoviciana Nutt. Gen. 2:20. 1818.

  Anemone nuttalliana DC. Syst. 1:193. 1818.

  Anemone patens L. var. *nuttalliana* (DC.) Gray, Man. ed. 5, 36. 1867.

  Pulsatilla ludoviciana (Nutt.) Heller, Cat. N. Am. Pl. ed. 2, 4. 1900.

Herbaceous perennial from a thick, brown caudex; basal leaves 2–3 ternately divided into linear divisions, densely villous, on villous petioles to 20 cm long; involucral leaves 3, similar in shape and pubescence, sessile; flowering stem silky-villous, to 40 cm tall; flower solitary, terminal, up to 8 cm broad, on a villous peduncle to 4 cm long; sepals 5–7, free, petallike, blue or whitish, ovate-oblong, to 4 cm long; petals absent; stamens numerous, some of them without anthers and glandular; pistils numerous, densely villous, with long, villous styles; achenes numerous in a head, fusiform, densely villous, the persistent plumose styles 2–4 cm long.

COMMON NAME: Pasque-flower.

HABITAT: Prairies; morainal hills.

RANGE: Michigan to Alaska, south to Washington, Texas, and northern Illinois.

ILLINOIS DISTRIBUTION: Confined to a few counties in extreme northern Illinois.

The pasque-flower is one of the most beautiful species of the dry prairies in northern Illinois. The blue or nearly whitish flowers may be as much as 8 cm broad. They open in March and April.

The achenes bear persistent, plumose styles, strongly resembling various species of *Clematis*.

The entire plant is silky-hairy.

Much controversy exists concerning the correct name for this plant. It belongs to section *Pulsatilla* of *Anemone*, differing from the other Anemones by the plumose styles and the presence of some antherless stamens.

62. *Anemone patens* (Pasque-flower). *a*. Habit, in flower, X½. *b*. Habit, in fruit,
X1. *c*. Flower, X¾.

63. *Anemone caroliniana* (Carolina Anemone). *a*. Habit, X1. *b*. Flower, X2½. *c*. Fruiting head, X2.

2. **Anemone caroliniana** Walt. Fl. Car. 157. 1788. *Fig. 63.*

Anemone tenella Pursh, Fl. Am. Sept. 2:386. 1814.

Anemone caroliniana Walt. f. *violacea* Clute, Am. Bot. 28:80. 1922.

Herbaceous perennial from a tuber up to 1 cm in diameter; basal leaves ternately divided, the divisions deeply toothed or lobed, sparsely pubescent, on slender petioles about as long as the leaves; involucral leaves usually 2, borne near the bottom of the flowering stem, ternately divided, the divisions not as incised as those of the basal leaves, sessile; flowering stem sparsely pubescent, to 25 cm tall; flower solitary, to 2 cm broad, terminal, on a sparsely pubescent peduncle; sepals 10–20, white or violent, linear-oblong, to 2 cm long; petals absent; stamens numerous, all anther-bearing; pistils numerous, densely pubescent; fruiting head ellipsoid, to 1.5 cm long, with numerous achenes, the achenes densely woolly.

COMMON NAME: Carolina Anemone.

HABITAT: Hill prairies; sandy pastures.

RANGE: Wisconsin to South Dakota, south to Texas and Florida.

ILLINOIS DISTRIBUTION: Confined to the northern half of the state.

The Carolina anemone is readily distinguished by its large, solitary flower with 10–20 petallike sepals. The involucral leaves are also borne lower on the flowering stem than in most of the other species of *Anemone*.

Although the usual habitat for this species in Illinois is in hill prairies, Swink (1974) reports a Grundy County station in sandy pastures, associated with *Androsace occidentalis* and *Draba reptans*, two of its associates on hill prairies.

This species blooms in April and May.

3. **Anemone canadensis** L. Syst. ed. 12, 3:App. 231. 1768. *Fig. 64.*

Anemone pensylvanica L. Mant. 2:247. 1771.

Herbaceous perennial from slender rhizomes; basal leaves palmately 3– to 7–divided, the division usually 3–cleft and coarsely toothed, conspicuously reticulate-veined, pilose on the veins beneath, to 15 cm long, about as broad, on long, sparsely hairy petioles; flowering stem stout, to 70 cm tall, sparsely pubescent, with

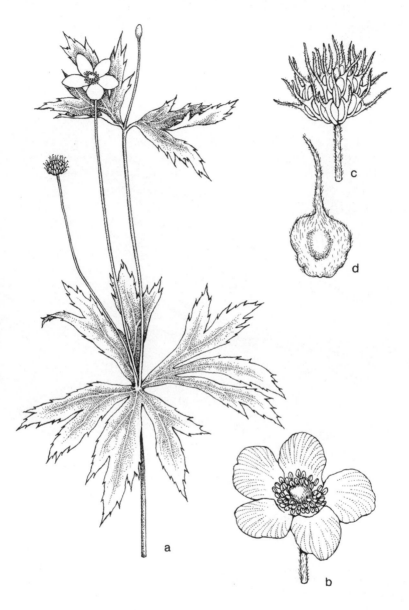

64. *Anemone canadensis* (Meadow Anemone). *a*. Upper part of plant, X½. *b*. Flower, X1½. *c*. Fruiting head, X2½. *d*. Achene, X4.

both primary and secondary involucres, the involucral leaves similar to the basal leaves, except sessile; flowers 1–6, to 3 cm broad, on a more or less pubescent peduncle; sepals 5, white, oblong to obovate, to 2.5 cm long; petals absent; stamens numerous, all anther-bearing; pistils numerous, pubescent; fruiting head globose, to 1.5 cm in diameter, with numerous achenes, the achenes nearly orbicular, subulate-beaked, pubescent, flat, up to 6 mm across, the beak 2–5 mm long.

COMMON NAME: Meadow Anemone.

HABITAT: Prairies; open woodlands.

RANGE: Nova Scotia to British Columbia, south to New Mexico, Illinois, and West Virginia.

ILLINOIS DISTRIBUTION: Occasional in the northern three-fifths of the state, rare elsewhere.

The meadow anemone is distinguished by its sessile involucral leaves and its subulate-beaked achenes.

Swink (1974) records its more favorable habitat in prairie remnants with somewhat calcareous soils.

The first collection from the southern one-fourth of the state was made from the edge of a woods in Riverside Park, Murphysboro (Jackson County). That station is no longer extant, but this species has recently been found in wet soil at the Greentree Reservoir in Jackson County.

Illinois botanists from Beck (1828) to Cowles (1901) called this species *Anemone pensylvanica* L., but this is clearly a synonym for *A. canadensis*.

4. **Anemone quinquefolia** L. Sp. Pl. 541. 1753. *Fig. 65.*

Anemone nemorosa L. var. *quinquefolia* (L.) Gray, Man. Bot., ed. 5, 38, 1867.

Anemone quinquefolia L. var. *interior* Fern. Rhodora 37:260. 1935.

Herbaceous perennial from slender white rhizomes; basal leaf one, appearing after the flowering stem elongates, on a long petiole, 3- to 5-foliolate, the leaflets oblong to elliptic, cuneate, glabrous, coarsely toothed; flowering stem slender, to 25 cm long, glabrous or villous, with one whorl of 3 involucral leaves on slender petioles to 2 cm long, the leaves shallowly or deeply 3- to 5-lobed or –divided, coarsely toothed, glabrous; flower solitary, to 2.5 cm broad,

65. *Anemone quinquefolia* (Wood Anemone). *a*. Habit, X½. *b*. Flower, X2.

on a slender, glabrous or villous peduncle to 3.5 cm long; sepals (4–) 5 (–9), white or purplish, oval to obovate, to 2.5 cm long; petals absent; stamens numerous, all anther-bearing; pistils numerous, pubescent; fruiting head globose, to 1.5 cm in diameter, with numerous achenes, the achenes oblong, hirsutulous, to 4.5 mm long, with a slender, curved beak 1–2 mm long.

COMMON NAME: Wood Anemone.

HABITAT: Moist woods; prairie remnants.

RANGE: Quebec to Manitoba, south to Iowa, Kentucky, and North Carolina.

ILLINOIS DISTRIBUTION: Confined to the northern one-fourth of the state; also Hardin and Menard counties.

The wood anemone occurs primarily in rich, mesic woods, although it sometimes is found in prairie remnants along with such species as *Comandra richardsiana, Zizia aurea, Heuchera richardsonii,* and *Dodecatheon media.*

All Illinois specimens have somewhat villous stems and have been designated var. *interior* by Fernald.

This species is recognized by its solitary basal leaf which develops after the flowering stalk elongates, the petiolate involucral leaves, and the hirsutulous achenes in a globose head.

The flowers of the wood anemone are produced in April and May.

5. **Anemone cylindrica** Gray, Ann. Lyc. N. Y. 3:221. 1836. *Fig. 66.*

Herbaceous perennial; basal leaves long-petiolate, deeply 3– to 5–parted, the divisions obovate to oblanceolate, cuneate, coarsely toothed, pubescent; flowering stem to 65 cm tall, silky-villous, with one whorl of 5–9 involucral leaves on petioles up to 2 cm long, the leaves similar to the basal leaves in shape and pubescence; flower one per peduncle, but sometimes several per plant, to 2 cm broad, on an elongated peduncle; sepals 5–6, greenish-white, oblong, to 1.5 cm long, silky on the outer surface; petals absent; stamens numerous, all anther-bearing; pistils numerous, densely pubescent; fruiting head elongate-cylindric, to 4 cm long, to 1 cm thick, with numerous achenes, the achenes fusiform, densely woolly, minutely beaked.

COMMON NAME: Thimbleweed.

HABITAT: Open woods; prairies; fields.

RANGE: Maine to Alberta, south to Arizona, central Illinois, and New Jersey.

ILLINOIS DISTRIBUTION: Confined mostly to the northern two-thirds of the state.

The greatly elongated, cylindrical fruiting head dis-

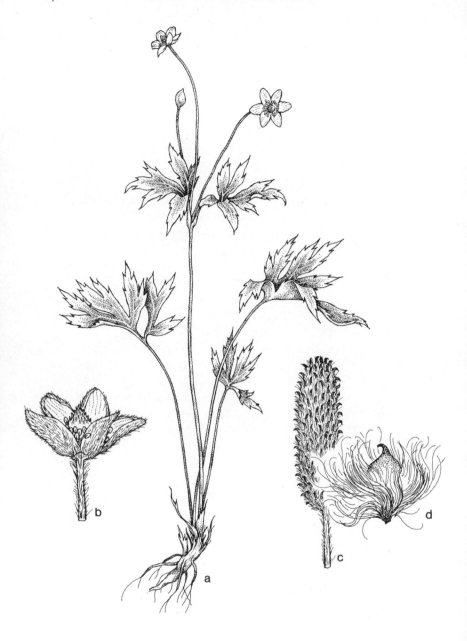

66. *Anemone cylindrica* (Thimbleweed).   *a*. Habit, X½.   *b*. Flower, X2.   *c*. Fruiting head, X1½.   *d*. Achene, X4.

tinguishes this species from all other anemones in Illinois. The presence of five or more involucral leaves on the flowering stem further differentiates this species from *A. virginiana*.

The thimbleweed, named from the shape of the fruiting head, flowers from May to August.

6. Anemone virginiana L. Sp. Pl. 540. 1753. *Fig. 67*.

Herbaceous perennial; basal leaves long-petiolate, deeply 3– to 5–parted, the divisions obovate to oblanceolate, cuneate, coarsely toothed, pubescent; flowering stem to 1 m tall, villous, with one whorl of 3 involucral leaves on petioles to 4 cm long, the leaves similar to the basal leaves in shape and pubescence; flower one per peduncle, but sometimes several per plant, to 2 cm broad, on an elongated peduncle; sepals 5 (–6), white or greenish-white, oblong, to 1.5 cm long, pubescent on the outer surface; petals absent; stamens numerous, all anther-bearing; pistils numerous, densely pubescent; fruiting head oblongoid-cylindric, to 3 cm long, to 1.5 cm thick, with numerous achenes, the achenes fusiform, densely woolly, minutely beaked.

COMMON NAME: Tall Anemone.

HABITAT: Woodlands.

RANGE: Maine to Minnesota, south to Kansas and Georgia.

ILLINOIS DISTRIBUTION: Common throughout the state.

This species is the commonest *Anemone* in the state, where it probably occurs in every county.

The tall anemone has a fruiting head never more than twice as long as thick, thus distinguishing it from the similar *A. cylindrica*.

This species flowers from late May to August.

13. *Myosurus* L.–*Mousetail*

Annual herbs from fibrous roots; leaves basal, linear, entire; flowering scape unbranched, 1-flowered; sepals 5, free, spurred at base; petals 5, free, or absent; stamens 5–20, free; pistils numerous, free, on an elongated receptacle, the ovaries superior, with 1 ovule; achenes borne on an elongated receptacle.

Myosurus is a genus of five species found in man parts of the world.

Only the following species occurs in Illinois.

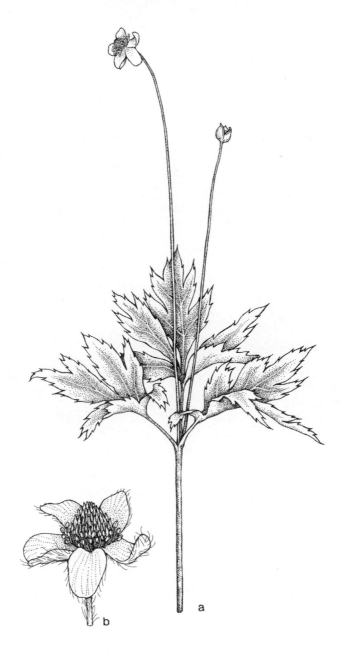

67. Anemone virginiana (Tall Anemone). *a*. Upper part of plant, X½. *b*. Flower, X2½.

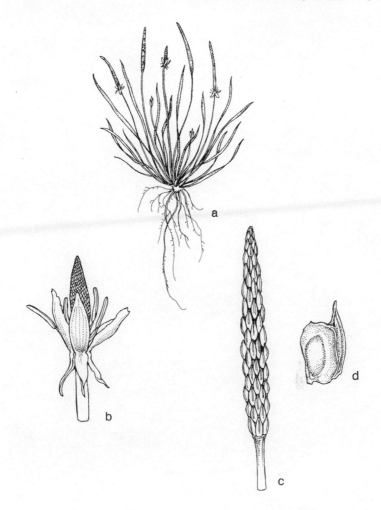

68. *Myosurus minimus* (Mousetail). *a*. Habit, X½. *b*. Flower, X4. *c*. Fruiting head, X2½. *d*. Achene, X12½.

1. Myosurus minimus L. Sp. Pl. 284. 1753. *Fig. 68*.

Annual from fibrous roots; all leaves basal, linear, to 9 cm long, to 5 mm broad, obtuse to subacute, glabrous; flowering scape to 15 cm tall, unbranched, glabrous, terminated by a single flower; sepals 5 (–7), greenish-yellow, elliptic, subacute, to 5 mm long, long-spurred at the base; petals 5, greenish-yellow, linear; stamens

10–18, free; pistils numerous on an elongated receptacle, glabrous; fruiting receptacle slender, to 5 cm long, with numerous achenes, the achenes quadrate, obtuse, apiculate, glabrous.

COMMON NAME: Mousetail.
HABITAT: Moist ground in woods and fields.
RANGE: Southern Ontario to British Columbia, south to California, Texas, and Florida; South America; Europe; Asia; North Africa; Australia; New Zealand.
ILLINOIS DISTRIBUTION: Occasional to common in the southern three-fifths of the state; also LaSalle County.

Mousetail is a sometimes overlooked species occurring in moist fields or along the edges of woods. The elongated fruiting receptacle may attain a length of 5 cm, but it usually is much shorter.

This species flowers from April to June.

14. Aquilegia L.–Columbine

Perennial herbs; leaves 2–3 ternately compound; flowers large, showy; sepals 5, free, petallike; petals 5, long-spurred at the base; stamens numerous, free, some of them often sterile; pistils 5, the ovaries superior, with many ovules; fruit a head of erect, many-seeded follicles.

Aquilegia is a genus of about fifty species and many cultivated varieties. About half of the species in the genus occur in the western United States.

One native and one escaped species occur in Illinois.

KEY TO THE SPECIES OF Aquilegia IN ILLINOIS

1. Spurs of flower straight; flowers red and yellow ___ 1. *A. canadensis*
1. Spurs of flower hooked; flowers blue, purple, pink, or white _____
_____ 2. *A. vulgaris*

1. Aquilegia canadensis L. Sp. Pl. 533. 1753. *Fig. 69.*

Aquilegia coccinea Small, Bull. N. Y. Bot. Gard. 1:280. 1899.
Aquilegia canadensis L. var. *coccinea* (Small) Munz, Gent. Herb. 7:120. 1946.

Herbaceous perennial from fibrous roots; stems erect, to 1 m tall, branched, glabrous or occasionally pubescent; lower leaves biternate, on long, slender petioles; leaflets thin, obtusely lobed, obovate, green above, paler beneath, usually glabrous; upper leaves

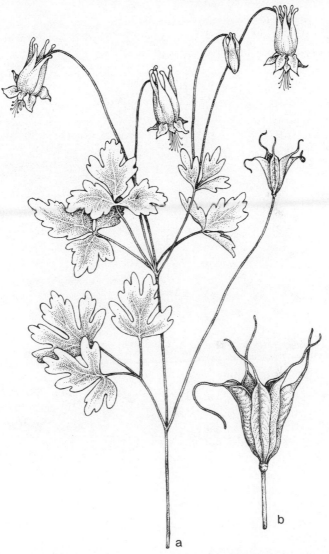

69. *Aquilegia canadensis* (Columbine). *a*. Upper part of plant, X½. *b*. Fruit, X2.

3–parted; flowers to 5 cm long, nodding; sepals 5, reddish-yellow, oblong to oblong-ovate, to 2 cm long, smaller than the petals; petals 5, bright red on the outside, yellow on the inside, prolonged backward into nearly straight, nectar-bearing spurs; stamens numerous, long-exserted; pistils 5, the styles exserted; fruiting head erect, with 5 follicles, the follicles to 2 cm long, slenderly beaked at the tip.

COMMON NAME: Columbine.

HABITAT: Usually rocky woods; dune woods; mesic forests; calcareous fens.

RANGE: Nova Scotia to Alberta, south to Texas and Florida.

ILLINOIS DISTRIBUTION: Occasional to common throughout the state.

The columbine is one of our most beautiful wildflowers. It occurs in a variety of woodland habitats, but usually in rocky woods.

Munz (1946) has divided this species into a number of varieties, based primarily on length and shape of sepals. He attributed var. *coccinea* (Small) Munz to Illinois, distinguishing it by longer sepals. Sepal size does not seem to be correlated with any other characters so that I feel there is no justification to maintain var. *coccinea*.

The columbine flowers from April to July.

2. Aquilegia vulgaris L. Sp. Pl. 533. 1753. *Fig. 70.*

Herbaceous perennial from fibrous roots; stems erect, to 70 cm tall, branched, glabrous or pubescent; lower leaves 2–3 ternate, on long, slender petioles; leaflets obtusely lobed, broadly obovate, green above, glaucous beneath, glabrous or pubescent; upper leaves 3–parted; flowers to 5 cm long, nodding; sepals 5, variously colored, oblong; petals 5, blue, purple, pink, or white, prolonged backward into stout, hooked, nectariferous spurs; stamens numerous, scarcely exserted; pistils 5, the styles scarcely exserted; fruiting head erect, with 5 follicles, the follicles to 1 cm long, slenderly beaked at the tip.

COMMON NAME: Garden Columbine.

HABITAT: Waste ground, where it has escaped from cultivation.

RANGE: Native of Europe; sparingly escaped in the northeastern United States.

ILLINOIS DISTRIBUTION: Collected from Champaign, Cook, DuPage, and Kankakee counties.

70. *Aquilegia vulgaris* (Garden Columbine).   *a*. Upper part of plant, X½.

The garden columbine, rarely escaped from gardens, occurs in a variety of flower colors. It flowers from May to July.

15. *Helleborus* L.–*Hellebore*

Perennial herbs; leaves palmately divided, mostly all basal; flowers large, solitary to several; sepals 5, free, petallike, usually becoming dry and persisting; petals 8–10, free, tubular, much shorter than the sepals; stamens numerous, free; pistils 3–10, developing into several-seeded follicles.

There are about fifteen species of *Helleborus*, all native to Europe and Asia. The best known species probably is *H. niger* L., the Christmas rose.

Only the following garden escape occurs in Illinois.

1. **Helleborus viridis** L. Sp. Pl. 558. 1753. *Fig. 71.*

Perennial herb from thick fibrous roots; basal leaves on petioles to 25 cm long, palmately 7– to 11–divided, to 30 cm broad, the divisions sharply serrate, oblong, acute, to 10 cm long, glabrous; upper leaves similar to the basal leaves but smaller and sessile; flowering stems stout, erect, glabrous, to 70 cm tall; flowers large, pendent, on glabrous peduncles; sepals 5, yellow-green, broadly oblong, obtuse, to 2 cm long; petals 8–10, tubular, 2-lipped, yellow-green, to 5 mm long; stamens numerous, longer than the petals; pistils 2–10, glabrous; follicles usually 2–5 (–10), glabrous, to 1.8 cm long, slenderly beaked.

COMMON NAME: Green Hellebore.

HABITAT: Woodlands (in Illinois).

RANGE: Native of Europe; rarely escaped from cultivation in North America.

ILLINOIS DISTRIBUTION: Clark, Richland, and Wabash counties.

The green hellebore is a coarse herb, occasionally grown in Illinois gardens, but rarely escaping. Robert Ridgway and Jacob Schneck, collecting in Richland and Wabash counties, respectively, in the 1870s, found it as an escape in woodlands.

The flowers bloom in May.

The white hellebore belongs to the lily family, while the helleborine is an orchid.

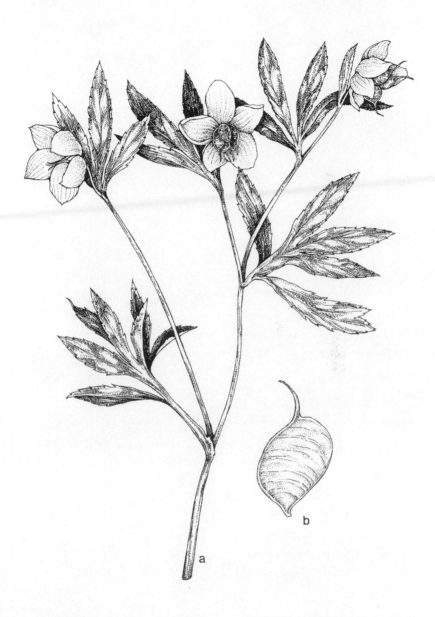

71. *Helleborus viridis* (Green Hellebore). *a*. Upper part of plant, in flower, X¾. *b*. Fruit, X1½.

16. Nigella L.–Fennel-flower

Annual herbs; leaves alternate, pinnately divided into filiform or linear segments; flowers showy, sometimes subtended by an involucre; sepals 5, free, petallike; petals 5, free, bifid; stamens numerous, free; pistils usually 5, united at least at the base; capsule dehiscent at the top, many-seeded.

Nigella is a genus of about twelve showy species, most of which may be grown as garden ornamentals. The genus is unique among the Ranunculaceae because of its pistils which are united at least at the base.

Only the following species has been found in Illinois.

1. Nigella damascena L. Sp. Pl. 584. 1753. *Fig. 72.*

Annual herb from fibrous roots; stems erect, much branched, glabrous, to 60 cm long; leaves alternate, deeply pinnately divided into many filiform segments, glabrous; flower solitary, to 3 cm across, subtended by an involucre similar in shape to the leaves; sepals 5, white or bluish, much longer than the petals; petals 5, white or bluish, tubular, 2–cleft at the apex; stamens numerous; pistils 5, united below, glabrous; capsules united nearly to the top, glabrous, with many black seeds.

COMMON NAME: Love-in-a-Mist.

HABITAT: Waste ground.

RANGE: Native of southern Europe; rarely escaped from cultivation in North America.

ILLINOIS DISTRIBUTION: I have seen one collection from Jackson County and one from Lake County.

The name love-in-a-mist alludes to the flower surrounded by the lacy involucre.

The species is remarkable among the Ranunculaceae by its united fruit.

The flowers occur from June to August.

17. Eranthis Salisb.–Winter Aconite

Perennial herbs; leaves palmately compound; flower solitary, large, showy; sepals 5 or more, free, petaloid; petals small, converted into 2-lipped nectaries; stamens numerous, free; pistils 3 or more, free; fruit a head of follicles.

The five species of this genus are native to Europe and Asia.

Only the following species has been found in Illinois.

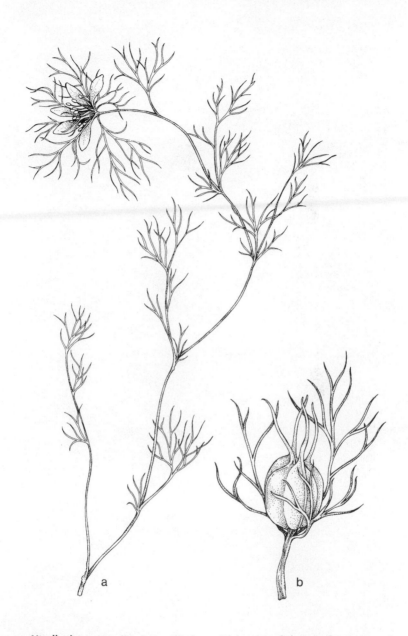

72. *Nigella damascena* (Love-in-a-Mist). *a*. Upper part of plant, in flower, X¾. *b*. Fruit, with bracts, X1½.

73. *Eranthis hyemalis* (Winter Aconite). *a*. Habit, in flower, X¾. *b*. Fruit, X4.

1. Eranthis hyemalis (L.) Salisb. Trans. Linn. Soc. 8:304. 1803.
Fig. 73.

Helleborus hyemalis L. Sp. Pl. 557. 1753.

Perennial herb from a bulbous base; stems erect, unbranched, glabrous, up to 20 cm tall; basal leaves 1–few, long-petiolate, up to 3
cm across, palmately divided into several bi- or trifid segments,
glabrous; cauline leaf 1, immediately subtending the flower, similar
to the basal leaves, but sessile; flower solitary, up to 3 cm across,
sessile; sepals 5–7 (–9), petaloid, membranous, obtuse, up to 1.5

cm long, yellow; petals 5 or more, reduced to bilobed nectaries, narrowed to the base, much smaller than the sepals; stamens numerous; pistils several; follicles usually 5 or more, flattened, up to 1 cm long, slenderly and excentrically beaked, several-seeded.

COMMON NAME: Winter Aconite.

HABITAT: Disturbed soil.

RANGE: Native of Europe; rarely escaped from cultivation.

ILLINOIS DISTRIBUTION: Known from a single collection in Belleville, St. Clair County.

The winter aconite is a garden escape in Illinois. It has been found in disturbed soil in Belleville where it has persisted for several years.

This genus is closely related to *Helleborus*, a genus which has persistent sepals and leafy stems.

The winter aconite flowers during February and March.

18. *Clematis* L.–*Clematis*

Climbing vines or erect perennial herbs; leaves opposite, simple or pinnately compound; flowers in cymose panicles or solitary, bisexual or unisexual; sepals 4–5, free, petallike; petals absent; stamens numerous, free; pistils numerous, free; fruit a head of achenes with persistent, plumose styles.

Clematis is a genus of about fifty species, occurring primarily in the Northern Hemisphere of both the Old and New Worlds.

Some botanists choose to segregate those erect species with solitary flowers into the genus *Viorna*.

Several species, varieties, and hybrids of *Clematis* are cultivated as garden ornamentals.

Because of its usually four valvate sepals, opposite leaves, and often plumose fruits, *Clematis* is generally segregated into tribe Clematideae of the Ranunculaceae.

KEY TO THE SPECIES OF Clematis IN ILLINOIS

1. Inflorescence paniculate; sepals thin, white; anthers blunt _____ 2
1. Flower solitary; sepals thick, bluish; anthers with an attenuated tip 3
   2. Leaves primarily 3-foliolate; anthers up to 1.5 mm long; achenes with spreading hairs _____ 1. *C. virginiana*
   2. Leaves primarily 5-foliolate; anthers at least 2 mm long; achenes with appressed, silky hairs _____ 2. *C. dioscoreifolia*

3. Tails of fruits densely plumose; only the tips of the sepals recurved _____ 3. *C. viorna*
3. Tails of fruits glabrous or pubescent, but not plumose; upper half of the sepals recurved _____ 4
    4. Leaves thick, conspicuously reticulate beneath; sepals less than 25 mm long, the margins not crisped _____ 5. *C. pitcheri*
    4. Leaves thin, not conspicuously reticulate beneath; sepals over 25 mm long, the margins crisped _____ 4. *C. crispa*

1. **Clematis virginiana** L. Amoen. Acad. 4:275. 1759. *Fig. 74.*

Dioecious, climbing perennial; stems usually glabrous, usually very much elongated; leaves trifoliolate, on glabrous petioles; leaflets thin, broadly ovate, acute at the apex, subcordate or rounded at the base, toothed or shallowly lobed, to 3 cm long, glabrous or nearly so; flowers numerous, in cymose panicles, creamy-white, to 3 cm wide; sepals 4 (–5), lanceolate, to 1.2 cm long; petals absent; stamens numerous (in the staminate flowers), the anthers up to 1.5 mm long; achenes (from the pistillate flowers) brownish, with spreading pubescence, the persistent styles plumose.

COMMON NAME: Virgin's Bower.
HABITAT: Woods and thickets.
RANGE: Nova Scotia to Manitoba, south to eastern Kansas, Louisiana, and Georgia.
ILLINOIS DISTRIBUTION: Occasional throughout the state.

    The virgin's bower is distinguished from all other members of the genus in Illinois except *C. dioscoreifolia* by its cymose-paniculate inflorescence. From *C. dioscoreifolia*, it differs by its essentially trifoliolate, thin leaves.

Variation occurs in the amount of pubescence (if any) on the leaflets.

    The virgin's bower appears to be most common in moist thickets where it sometimes runs rampant over low vegetation.

    This species flowers from July to early October.

2. **Clematis dioscoreifolia** Lévl. & Vaniot, Rep. Sp. Nov. 7. 339. 1909. *Fig. 75.*

Dioecious, climbing perennial; stems glabrous, much elongated; leaves mostly 5-foliolate, on glabrous petioles; leaflets rather thick,

74. *Clematis virginiana* (Virgin's Bower). *a*. Habit, X½. *b*. Flower, 2½. *c*. Flower, X2½. *d*. Achene, X5.

broadly ovate to suborbicular, obtuse at the apex, cordate at the base, crenate or entire, to 3 cm long, glabrous; flowers numerous, in cymose panicles, white, to 3 cm wide; sepals 4 (–5), lanceolate, to 1.7 cm long; petals absent; stamens numerous (in the staminate

75. *Clematis dioscoreifolia* (Virgin's Bower).   *a*. Habit, X½.   *b*. Flower, X2½.   *c*. Achene, X5.

flowers), the anthers over 1.5 mm long; achenes (from the pistillate flowers) pale brown, with appressed silky hairs, the persistent styles plumose.

COMMON NAME: Virgin's Bower.
HABITAT: Roadsides and borders of woods.
RANGE: Native of Japan and Korea; occasionally adventive in the eastern United States.
ILLINOIS DISTRIBUTION: Occasional in the state.

Very similar in appearance to the more common and native *C. virginiana*, this species is distinguished by its thicker, usually 5-foliolate leaves, its longer anthers, and its silky achenes.

It flowers from July to October.

3. **Clematis viorna** L. Sp. Pl. 543. 1753. *Fig. 76.*

Viorna viorna (L.) Small, Fl. S.E.U.S. 439. 1903.

Viorna·ridgwayi Standl. Smithson. Misc. Coll. 56:2, pl. 1. 1912.

Perennial climbing herb; stems 6-angled, glabrous except at the nodes, to 4 m long; leaves pinnately compound, or sometimes simple on the upper branches; leaflets rather thin, ovate to lanceolate to broadly elliptic, acute to acuminate, cuneate to rounded at the base, entire, glabrous above, puberulent beneath; peduncles usually with 2 bracts and 1 nodding flower; sepals 4, forming a campanulate calyx, very thick, purple, oblong-lanceolate, to 2.5 cm long, the tips acuminate, recurved; petals absent; stamens numerous, the anthers and filaments about equal in length; achenes ovoid, brownish, strigose, the persistent style plumose, to 2.5 cm long.

COMMON NAME: Leatherflower.
HABITAT: Along streams.
RANGE: Pennsylvania to Missouri, south to Texas and Georgia.
ILLINOIS DISTRIBUTION: Known first from Richland County, where it was collected by Robert Ridgway, near Olney, on June 6, 1910. It has subsequently been found in Jasper and Johnson counties.

The leatherflower gets its name from the thick, leathery sepals which form a bell-shaped flower. This species differs from the other leathery-flowered species of *Clematis* in Illinois by its plumose styles.

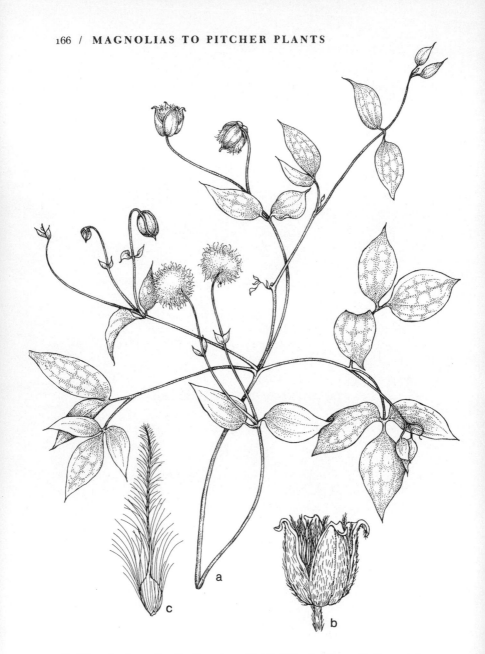

76. *Clematis viorna* (Leatherflower). *a*. Habit, X½. *b*. Flower, X1½. *c*. Achene, X5.

The first Illinois collection, made by Ridgway in 1910, differs somewhat from other specimens of *C. viorna* by its acuminate leaflets. Standley judged this character to be important enough to designate the Ridgway collection a new species, *C. ridgwayi* Standl. Since there are no other differences exhibited by *C. ridgwayi*, there is no justification for maintaining it.

The leatherflower blooms from May to July.

4. Clematis crispa L. Sp. Pl. 543. 1753. *Fig. 77.*

Clematis simsii Sweet, Hort. Brit. 1. 1826.

Viorna crispa (L.) Small, Fl. S.E.U.S. 437. 1903.

Perennial climbing herb; stems more or less terete, glabrous, to 4 m long; leaves 3– to 9–foliolate, rarely simple on the upper branches; leaflets rather thin, entire or 3-lobed, ovate to lanceolate, acute, subcordate to cuneate at the base, not prominently reticulate, glabrous; peduncles usually bractless, with 1 nodding flower; sepals 4, forming a campanulate calyx, rather thin, bluish-purple, oblong-lanceolate, undulate, 2.5–4.0 cm long, acuminate, recurved in their upper half; petals absent; stamens numerous, the anthers and filaments about equal in length; achenes ovoid, brownish, strigose, the persistent styles silky but not plumose, to 2.5 cm long.

COMMON NAME: Blue Jasmine.

HABITAT: Low woods.

RANGE: Virginia to southern Missouri, south to Texas and Florida.

ILLINOIS DISTRIBUTION: Confined to the southern one-third of the state.

This southern species is the only solitary-flowered *Clematis* in Illinois with thin sepals. It is similar to *C. pitcheri* because of its nonplumose persistent styles, but *C. pitcheri* has leathery sepals.

The first Illinois collection was made by George Engelmann in June, 1824, along Silver Creek, in St. Clair County. Today this species is known in Illinois only from the southernmost tip of the state, where it occurs in cypress swamps or in areas which have been cypress swamps.

The blue jasmine flowers from April to July.

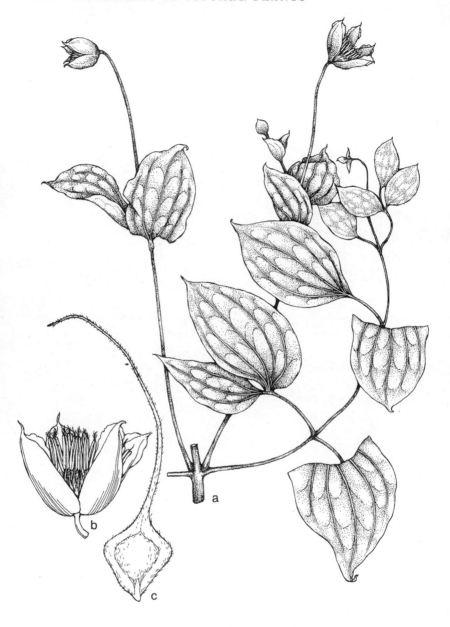

77. *Clematis crispa* (Blue Jasmine).   *a*. Habit, X⅓.   *b*. Flower, X1.   *c*. Fruit, X7½.

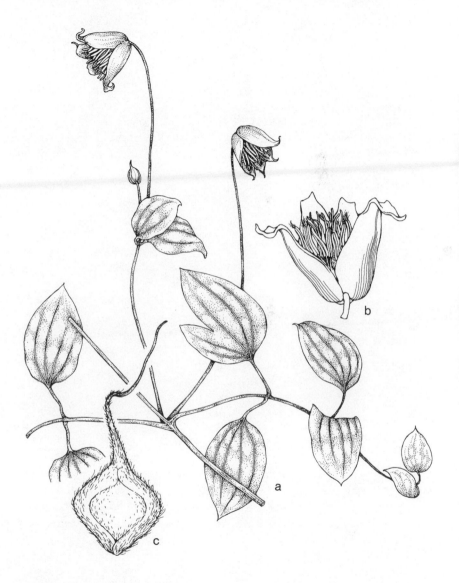

78. *Clematis pitcheri* (Leatherflower).   *a*. Habit, X½.   *b*. Flower, X1½.   *c*. Achene, X7½.

5. **Clematis pitcheri** Torr. & Gray, Fl. N. Am. 1:10. 1838. *Fig. 78.*

Viorna pitcheri (Torr. & Gray) Britt. in Britt. & Brown, Ill. Fl. N.E. States 2:123. 1913.

Perennial climbing herb; stems more or less terete, glabrous, to 4 m long; leaves 3– to 9–foliolate, rarely simple on the upper branches; leaflets thick, entire or 3–lobed, ovate, acute at the tip, rounded or cordate at the base, prominently reticulate, glabrous; peduncles usually bractless, with 1 nodding flower; sepals 4, forming a campanulate calyx, thick, purple, oblong-lanceolate, to 2.5 cm long, acuminate, recurved in the upper one-third; petals absent; stamens numerous, the anthers and filaments about equal in length; achenes ovoid, brownish, short-pubescent, the persistent styles glabrous or pilose but not plumose, to 2.5 cm long.

COMMON NAME: Leatherflower.

HABITAT: Woods and thickets.

RANGE: Indiana to Nebraska, south to Texas and Tennessee.

ILLINOIS DISTRIBUTION: Throughout the state, except for the extreme northern counties.

This leatherflower differs from *C. crispa* by its leathery sepals and from *C. viorna* by its eplumose persistent styles. The leaflets are more prominently reticulate than in any of our other species of *Clematis*.

It flowers from May to September.

BERBERIDACEAE–BARBERRY FAMILY

Shrubs or herbs, sometimes with prickly stems; leaves alternate or basal, simple or compound; stipules present or absent; flowers solitary or in racemes; sepals 4–9, free, usually petallike and caducous; petals 6–9, free; stamens 6–18, free; pistil one, the ovary superior, with 2–several ovules; fruit a berry or capsule.

The barberry family is composed of about ten genera and one hundred and fifty species, mostly in temperate regions of the world.

Two distinctive growth forms are found in the Berberidaceae. There are prickly shrubs, represented by *Berberis* and the cultivated genus *Mahonia*, and unarmed herbs, such as *Podophyllum*, *Caulophyllum*, and *Jeffersonia*.

KEY TO THE GENERA OF Berberidaceae IN ILLINOIS

1. Prickly shrubs; fruit a red berry _____ 1. *Berberis*
1. Unarmed herbs; fruit a capsule or a blue or yellow berry _____ 2
   2. Flower solitary on each plant, white; leaves simple _____ 3
   2. Flowers in panicles, yellow-green; leaves ternately compound _____ 4. *Caulophyllum*
3. Leaves 7– to 9–lobed; sepals usually 6; stamens twice as many as the petals; fruit a berry _____ 2. *Podophyllum*
3. Leaves 2–lobed; sepals usually 4; stamens the same number as the petals; fruit a capsule _____ 3. *Jeffersonia*

1. *Berberis* L.–*Barberry*

Usually prickly shrubs with yellow wood; leaves simple (in the Illinois species); flowers solitary or in racemes; sepals 6–9, free, petallike, with 2–6 bracteoles at base; petals 6, free, each with a pair of basal glands; stamens 6, free, the anthers dehiscing by terminal pores; pistil one, the ovary superior, the stigma peltate; fruit a red berry, with 1–few seeds.

Most of the nearly one hundred species of *Berberis* in the world are native to Europe, northern Africa, and South America. Several species are widely cultivated as ornamentals in Illinois.

The mature stamens, when touched, tend to snap shut around the stigma.

KEY TO THE SPECIES OF Berberis IN ILLINOIS

1. Leaves entire; flowers solitary or in clusters of 2–4; prickles unbranched; berries dry _____ 1. *B. thunbergii*
1. Leaves toothed; flowers in racemes; prickles mostly forked; berries fleshy _____ 2
   2. Leaves with up to 20 teeth on the margin; branchlets brown; petals notched; berries ovoid _____ 2. *B. canadensis*
   2. Leaves with 25 or more teeth on the margin; branchlets gray; petals entire; berries oblongoid to ellipsoid _____ 3. *B. vulgaris*

1. Berberis thunbergii DC. Reg. Veg. Syst. 2:9. 1821. *Fig.* 79.
Low, densely branched shrub to 1.5 m tall; stems brown, glabrous, deeply grooved, with unbranched spines; leaves spatulate to obovate, to 3 cm long, glabrous or nearly so, without teeth; flowers solitary or in clusters of 2–4 from the axils of most of the leaves; sepals mostly 6, pale yellow; petals 6, in 2 series, pale yellow; berries red, ellipsoid to globose, dry.

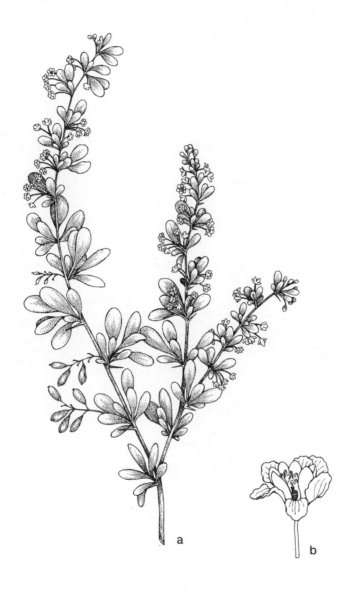

79. *Berberis thunbergii* (Japanese Barberry). *a*. Habit, X½. *b*. Flower, X5.

COMMON NAME: Japanese Barberry.

HABITAT: Thickets and waste areas.

RANGE: Native of Asia; often planted in Illinois and occasionally escaped.

ILLINOIS DISTRIBUTION: Scattered throughout Illinois.

The Japanese barberry is frequently grown as a hedge in Illinois. The red fruits, which persist on the plants throughout the winter months, are eaten by birds and the seeds thereby distributed in fields and woodlands throughout Illinois.

The entire leaves, unbranched spines, and dry fruits readily distinguish this species.

The flowers open in April and May.

2. **Berberis canadensis** Mill. Gard. Dict. ed. 8, no. 2. 1768.
   Fig. 80.

Berberis vulgaris L. var. *canadensis* (Mill.) Ait. Hort. Kew. 1:479. 1789.

Shrub to 2 m tall; stems dark brown, glabrous, grooved, with 3–pronged axillary spines; leaves spatulate to obovate, to 4 cm long, bristly serrate with up to 20 teeth on the margin; flowers few in racemes; sepals 6, free, yellow; petals 6, in 2 series, free, obovate, notched at the tip, yellow; berries red, fleshy, ovoid, 7–8 mm long.

COMMON NAME: American Barberry.

HABITAT: Moist soil; cliffs.

RANGE: West Virginia to southern Missouri, south to Georgia and Virginia.

ILLINOIS DISTRIBUTION: Tazewell Co.: Spring Lake, October 21, 1924, *G. C. Curran s.n.*; Jackson Co.: Fountain Bluff, May, 1973, *K. Wilson.*

This rare species was first collected in Illinois by G. C. Curran in 1924 in moist soil from Spring Lake, Tazewell County. During 1973, Wilson found it growing precariously from sandstone cliffs at Fountain Bluff in Jackson County.

Although Aiton and others have considered this taxon a variety of *B. vulgaris*, it differs from *B. vulgaris* by its brown twigs, fewer teeth on the leaves, and fewer flowers in the racemes.

The American barberry flowers in May.

80. *Berberis canadensis* (American Barberry). *a*. Habit, in flower, X½. *b*. Habit, in fruit, X½. *c*. Flower, X5.

3. Berberis vulgaris L. Sp. Pl. 330. 1753. *Fig. 81*.

Shrub to 2.5 m tall; stems gray, glabrous, grooved, with 3–pronged axillary spines; leaves spatulate to obovate, to 4.5 cm long, bristly serrate with 25 or more teeth on the margin; flowers several in racemes; sepals 6, free, yellow; petals 6, in 2 series, free, obovate,

81. *Berberis vulgaris* (Common Barberry). *a*. Habit, X½. *b*. Twig, with spines, X½. *c*. Flowering branchlet, X½. *d*. Flower, X5. *e*. Fruiting branchlet, X½.

entire, yellow; berries red, fleshy, oblongoid to ellipsoid, 8–12 mm long.

COMMON NAME: Common Barberry.

HABITAT: Along roads and other disturbed areas.

RANGE: Native of Europe and Asia; occasionally escaped throughout much of the eastern United States.

ILLINOIS DISTRIBUTION: At one time scattered throughout most of Illinois, but now much less common.

Because the common barberry serves as one of the hosts for the stem rust of wheat, much of it has been ‹eradicated from Illinois through a systematic program of extermination.

There are many cultivated variants of the common barberry, with the dark purple-leaved one probably the most commonly encountered.

This species flowers in May and June.

2. *Podophyllum* L.–*Mayapple*

Perennial herbs from widely creeping rhizomes and thick fibrous roots; aerial stems erect, with 1–2 leaves; leaves peltate, palmately lobed; flower solitary, axillary; sepals 6, free, petallike, caducous, subtended by 3 (–6) bracteoles; petals 6 (–9), free, longer than the sepals; stamens 12–18, free, the anthers dehiscing longitudinally; pistil one, the ovary superior, with many ovules; fruit a berry.

Podophyllum is a genus of four species, all but *P. peltatum* restricted to Asia.

Only the following species occurs in Illinois.

1. **Podophyllum peltatum** L. Sp. Pl. 505. 1753. *Fig. 82.*

Podophyllum peltatum L. f. *polycarpum* Clute, Am. Bot. 21:93. 1915.

Podophyllum peltatum L. f. *aphyllum* Plitt, Rhodora 33:229. 1931.

Podophyllum peltatum L. f. *deamii* Raymond, Rhodora 50:18. 1948.

Podophyllum peltatum L. f. *biltmoreanum* Steyerm. Rhodora 54:134. 1952.

Perennial herb from much-branched, stout rhizomes; stems to 75 cm tall, glabrous, bearing 1–2 leaves, or rarely leafless; leaves pel-

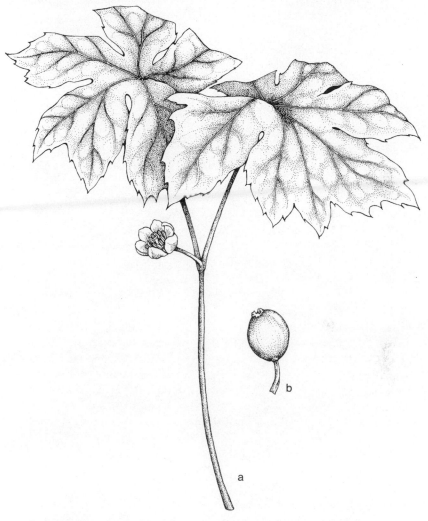

82. *Podophyllum peltatum* (Mayapple). *a*. Habit, X.2¼. *b*. Fruit, X.30.

tate, palmately 3– to 9–lobed, dentate, dark green and glabrous above, paler and glabrous or puberulent below, sometimes as much as 30 cm across; petioles long, glabrous; flower solitary, arising from the axil between the base of two leaves, the stout peduncle to 5 cm long, glabrous, yellow-green, rarely purplish; flowers nodding, white or rarely pinkish, to 4 cm across; sepals 6, free, petallike, caducous; petals 6 (–9), free, waxy, to 4 cm long; stamens 12–18, free; ovary one, rarely 2–8, yellow-green or rarely purple; fruit yellow to yellow-green to orange to purple, ovoid, to 5 cm long, glabrous, with numerous yellow or rarely dark purple-brown, arillate seeds.

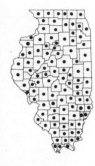

COMMON NAME: Mayapple.
HABITAT: Moist or dry woods.
RANGE: Quebec to southern Ontario and Minnesota, south to Texas and Florida.
ILLINOIS DISTRIBUTION: Common; in every county.

Mayapple, with its umbrellalike leaves and large, waxy flowers, is one of the most familiar wild flowers in Illinois. It is a somewhat variable species, but most of the variations are rarely encountered.

Specimens with more than one ovary per flower, designated f. *polycarpum* Clute, have been found in Lake County. Others without any leaves, known as f. *aphyllum* Plitt, are recorded from Pope and Will counties. Forma *biltmoreanum* Steyerm., with orange-colored fruits, was described from plants collected in Lake County. Forma *deamii* Raymond, from Cook County, has pink petals, purple ovaries and fruits, purple peduncles, and dark purple-brown seeds.

Mayapple frequently can be found in large, dense colonies. In such colonies, it is common for all of the aerial plants to be connected to the same highly branched rhizome.

Some three-lobed immature leaves may resemble leaves of *Jeffersonia diphylla*.

The rhizomes as well as the leaves are poisonous, although the rhizomes do have cathartic properties. On the other hand, the fruits are edible, having a slightly sweet, acidic taste. In earlier days, it was made into jellies or preserves.

The flowers appear in April and May. The fruits ripen in July and August.

3. *Jeffersonia Bart.* – Twinleaf

Perennial herbs from fibrous roots; leaves basal, long-petiolate, palmately lobed; flower solitary, on slender scapes; sepals 4, free, petallike, caducous; petals 8, free, longer than the sepals; stamens 8,

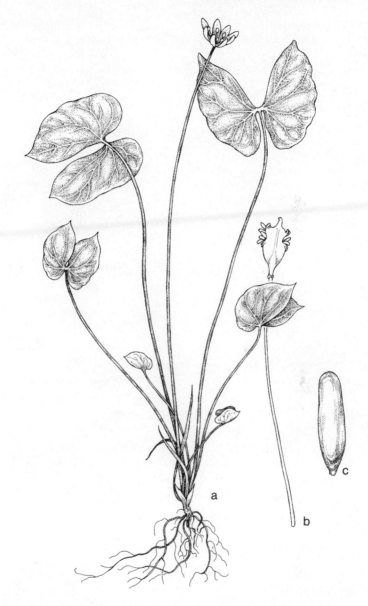

83. Jeffersonia diphylla (Twinleaf). *a*. Habit, in flower, X¼. *b*. Fruit, X½. *c*. Seed, X5.

free, the anthers dehiscing by terminal valves; pistil one, the ovary superior, with many ovules; fruit a leathery capsule, with arillate seeds.

In addition to our species, there is a second species in Manchuria.

Only the following species occurs in Illinois.

1. **Jeffersonia diphylla** (L.) Pers. Syn. 1:418. 1805. *Fig. 83.*

Podophyllum diphyllum L. Sp. Pl. 505. 1753.

Perennial herb from fibrous roots; leaves divided into two reniform-ovate leaflets, the leaflets to 15 cm long, to 4 cm broad, glabrous, glaucous below, entire or sinuate-dentate; flowering scape to 20 cm long at anthesis, to 45 cm long in fruit, glabrous; flower solitary, white, up to 2.2 cm across; sepals 4, free, white, caducous; petals 8, white, oblong, to 1.5 cm long; stamens 8, free; capsule leathery, pyriform, to 2 cm long, opening halfway around near the top, forming a lid.

COMMON NAME: Twinleaf.

HABITAT: Moist woods.

RANGE: New York to Wisconsin, south to Iowa, Illinois, and Virginia.

ILLINOIS DISTRIBUTION: Occasional in the northern three-fourths of Illinois; also Saline County.

Twinleaf occurs in rich woods, often on north-facing slopes. Its southernmost station is at Stone Face, in Saline County.

Reports from Jackson, Massac, Pope, and Pulaski counties by Jones et al. (1955) are apparently based on immature 3-lobed-leaf specimens of *Podophyllum peltatum.*

Twinleaf flowers in April and May.

4. *Caulophyllum Michx.*–Blue Cohosh

Perennial herbs with knotty rhizomes and matted fibrous roots; leaves ternately compound; inflorescence a terminal raceme or panicle; sepals 6, free, petallike, caducous, subtended by 3–4 bracteoles; petals 6, free, shorter than the sepals; stamens 6, free, the anthers dehiscing by terminal valves; pistil one, the ovary superior, with 2 ovules; fruit a blue berry.

Caulophyllum is composed of two species, ours and a second one in eastern Asia.

Only the following species occurs in Illinois.

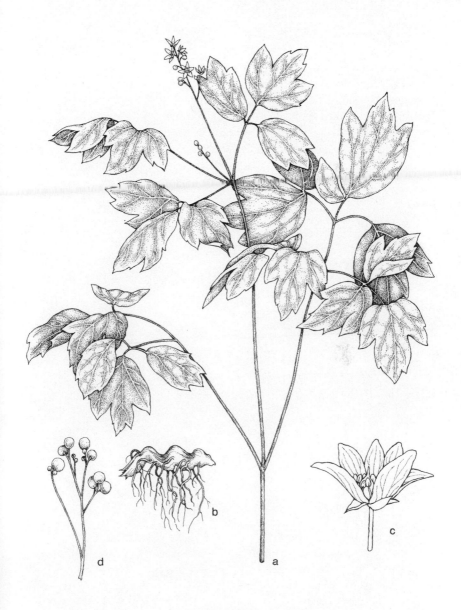

84. *Caulophyllum thalictroides* (Blue Cohosh). *a*. Upper part of plant, in flower, X.2¼. *b*. Rhizome, X.2¼. *c*. Flower, X2¼. *d*. Cluster of fruits, X.30.

1. **Caulophyllum thalictroides** (L.) Michx. Fl. Bor. Am. 1:205. 1803. *Fig. 84.*

Leontice thalictroides L. Sp. Pl. 312. 1753.

Perennial herb from knotty rhizomes; stems to 1 m tall, glabrous, sometimes glaucous, bearing a few leafless sheaths near the base and one ternately compound leaf near the tip; leaf triternately compound, sessile or nearly so, the leaflets oval, oblong, or obovate, 3- to 5-lobed, glabrous, glaucous, to 6 cm long; inflorescence a terminal panicle, many-flowered; flowers appearing when the leaves are less than half expanded, greenish-yellow or greenish-purple, to 1.2 cm across; sepals 6, free, elliptic, subacute, to 8 mm long, caducous; petals 6, free, very small, thick, glandlike; stamens 6, free; berries blue, globose, to 8 mm in diameter, glabrous, glaucous, on stout stalks to 6 mm long.

COMMON NAME: Blue Cohosh.

HABITAT: Moist woods.

RANGE: New Brunswick to Manitoba and North Dakota, south to Nebraska, Tennessee, and South Carolina.

ILLINOIS DISTRIBUTION: Occasional throughout the state.

Blue cohosh has leaves reminiscent of those found in the ranunculaceous genus *Thalictrum*, hence the specific epithet.

The flowers, which appear in April and May, open when the leaves are less than half expanded.

The so-called berries are actually seeds with fleshy blue integuments, becoming exposed when the membranous pericarp falls away early.

PAPAVERACEAE–POPPY FAMILY

Herbs with milky, colored, or clear sap; leaves alternate or basal, usually lobed or dissected; flowers actinomorphic or zygomorphic, variously arranged; sepals 2, free; petals 4–6, free or united in pairs, with or without a spur at the base; stamens 6 or numerous; pistil one, the ovary superior, the stigmas 2-lobed or 2-horned or united into a radiating disc; fruit a capsule, dehiscing longitudinally or by terminal pores or valves, or fruit 1-seeded and indehiscent.

As treated here, the often segregated Fumariaceae are included within the Papaveraceae. This is in accordance with the system of classification proposed by Thorne in 1968. The decision to combine the two families here is an arbitrary one, since there are perhaps just as many arguments for keeping the two families separate.

The basic differences separating these two groups are actinomorphic vs. zygomorphic flowers, stamens numerous vs. stamens six, and sap milky or colored vs. sap clear.

Most botanists will agree that the two groups are closely related, and that most members of the poppy group are more primitive than most members of the fumitory group.

With the two families combined, the Papaveraceae are composed of about fifty genera and seven hundred species, distributed primarily in temperate and subtropical regions of the world.

KEY TO THE GENERA OF Papaveraceae IN ILLINOIS

1. Flowers actinomorphic; stamens numerous; sap milky or variously colored _____ 2
1. Flowers zygomorphic; stamens 6; sap clear _____ 10
   2. Flowers basically white (pink in a rare form of *Sanguinaria*) ___ 3
   2. Flowers not white _____ 6
3. Perianth parts 6 or more _____ 4
3. Perianth parts 2 _____ 3. *Macleaya*
   4. Plants without stems; petals 6 or more _____ 1. *Sanguinaria*
   4. Plants with stems; petals 4–6 _____ 5
5. Leaves prickly _____ 2. *Argemone*
5. Leaves not prickly _____ 4. *Papaver*
   6. Leaves pinnatifid _____ 7
   6. Leaves ternately divided _____ 7. *Eschscholtzia*
7. Leaves prickly _____ 2. *Argemone*
7. Leaves not prickly _____ 8
   8. Flowers red or orange; sap milky; capsule opening by pores along the edge _____ 4. *Papaver*
   8. Flowers yellow; sap yellow; capsule opening from the bottom upward _____ 9
9. Capsule bristly; style distinct; petals 2–3 cm long __ 5. *Stylophorum*
9. Capsule smooth; style inconspicuous or absent; petals about 1 cm long _____ 6. *Chelidonium*
   10. Corolla with two spurs, or subcordate _____ 11
   10. Corolla with a single spur _____ 12
11. Leaves all from the base; plants erect _____ 8. *Dicentra*
11. Leaves cauline and alternate; plants climbing _____ 9. *Adlumia*

12. Flowers yellow or pink; capsule elongated, several-
seeded _____ 10. *Corydalis*
12. Flowers purplish, tipped with red; fruit nearly round, 1-
seeded _____ 11. *Fumaria*

1. Sanguinaria L.–Bloodroot

Herbaceous perennial from a thick rhizome with red sap; leaves basal, palmately lobed; flower solitary on a scape; sepals 2, free, caducous; petals 8–16, free; stamens numerous, free; ovary superior, with numerous ovules; fruit a capsule, with several smooth seeds.

Only the following species comprises the genus.

1. Sanguinaria canadensis L. Sp. Pl. 505. 1753. *Fig. 85.*

Sanguinaria rotundifolia Greene, Pittonia 5:35. 1905.
Sanguinaria canadensis L. var. *rotundifolia* (Greene) Fedde, Pflanzenr. 104:25. 1909.
Sanguinaria canadensis L. f. *colbyorum* Benke, Rhodora 35:45. 1933.

Rhizome thick, bearing numerous fibrous roots; leaves palmately 3- to 9-lobed, the margins dentate or undulate, membranaceous or subcoriaceous, more or less glaucous, glabrous, to 25 cm broad at maturity; petioles to 40 cm long, glabrous; flowering scape to 45 cm long, glabrous, sometimes shorter than or sometimes longer than the petioles; flowers to 3.2 cm broad, white, rarely pink; sepals 2, free, caducous; petals 8–16, oblong to spatulate, falling after one day or two; capsule ellipsoid to fusiform, 2-valved, to 2.5 cm long, containing several smooth, crested seeds.

COMMON NAME: Bloodroot.

HABITAT: Rich woods.

RANGE: Quebec to Manitoba and North Dakota, south to Louisiana and Florida.

ILLINOIS DISTRIBUTION: Common throughout the state.

This attractive wild flower is one of the first to bloom in Illinois, coming into flower in early March. The petals of each flower usually fall away after one day.

Distinguishing features of the bloodroot are the red sap in the rhizome, the palmately lobed basal leaves, and the flower composed of two sepals, 8–16 petals, and numerous stamens.

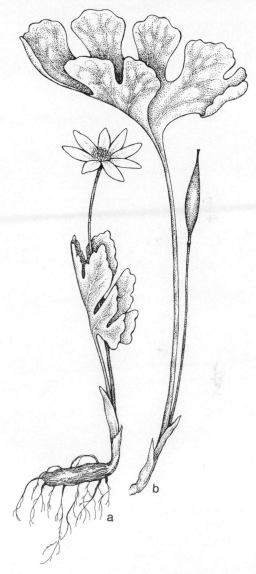

85. *Sanguinaria canadensis* (Bloodroot). *a*. Habit, in flower, X½. *b*. Habit, in fruit, X½.

Forma *colbyorum* Benke, a form with pink petals, was described from a collection made at Crystal Lake, McHenry County, on May 1, 1932, by E. H. Colby.

Although some botanists divide our species into var. *canadensis* and var. *rotundifolia*, on the basis of leaf texture and dentation, I am unable to differentiate among most Illinois material.

The red rhizome has been used as an emetic and purgative. It has a poisonous effect if used excessively.

2. *Argemone* L.–*Prickly Poppy*

Herbaceous annuals or biennials, with yellow sap; leaves cauline, alternate, pinnatifid, spiny-toothed; flowers showy, usually pedunculate; sepals 2–3, free, often prickly; petals 4–6, free; stamens numerous, free, the anthers dehiscing longitudinally; pistil one, the ovary superior, the dilated stigma 3- to 6-radiate; fruit a prickly capsule, dehiscing at the apex, with numerous seeds.

Argemone is a genus of about a dozen species, native to North America, particularly in the western United States. It is related to the genus *Papaver* by virtue of its apical capsule dehiscence. It differs by its spiny-toothed leaves.

A revision of the genus for North America has been prepared by Ownbey (1958).

KEY TO THE SPECIES OF Argemone IN ILINOIS

1. Petals yellow or orange or cream; leaves of a uniform color; stamens 30–50 _____ 1. *A. mexicana*
1. Petals white or pink; leaves with patches of pale green; stamens 50 or more _____ 2. *A. albiflora*

1. **Argemone mexicana** L. Sp. Pl. 1:508. 1753. *Fig.* 86.

Stout annual; stems to 80 (–100) cm tall, often branched from near the base, sparsely prickly; leaves glaucous, the light blue markings over the veins conspicuous, the lower leaves oblanceolate, the middle and upper leaves becoming broadly elliptic to ovate, clasping, lobed halfway or more to the middle, the lobes oblong, the marginal teeth spinose, the lower surface sparsely prickly on the main veins; flowers 4–7 cm in diameter, subtended by 1–2 bracts; petals yellow or orange or cream, the outer obovate, the inner obovate to obcuneate; stamens 30–50, yellow; stigma purple, about 1.5–4.0 mm. wide; capsule 4- to 6-carpellate, oblong to broadly elliptic, with spines up to 10 mm long; seeds 1.6–2.0 mm. long.

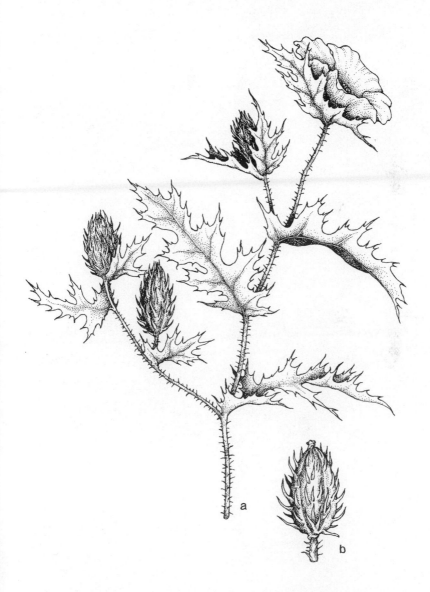

86. *Argemone mexicana* (Prickly Poppy).   *a*. Habit, X½.   *b*. Fruit, X1.

87. Argemone albiflora (White Prickly Poppy). *a*. Habit, X½. *b*. Fruit, X1.

COMMON NAME: Prickly Poppy.

HABITAT: Waste ground.

RANGE: Native of Mexico and the southwestern United States; occasionally adventive eastward.

ILLINOIS DISTRIBUTION: Adventive in Henderson, Macon, Menard, and Stephenson counties.

This native southwestern species has large, yellow flowers with 30–50 stamens. It is less commonly found in Illinois than A. *albiflora*.

Prickly poppy flowers from June to September.

2. Argemone albiflora Hornem. Hort. Haf. 489. 1815. *Fig. 87*,

Argemone alba Lestib. Bot. Belg. ed. 2 (3), 2:132. 1799, *nomen nudum*, not A. *alba* Jones (1823).

Stout annual or biennial; stems to 100 cm tall, sometimes branched from near the base, sparsely prickly; leaves uniformly green, the basal and lower cauline leaves lobed ½–⅔ or more to the middle, the middle and upper leaves less deeply lobed, glabrous; flowers 5–10 cm in diameter, subtended by 1–2 bracts; petals white or pink, the outer suborbicular, the inner obovate to obcuneate; stamens 150 or more; stigma purple, 3–4 mm wide; capsule 4- to 7-carpellate, oblong to oblong-elliptic, with spines up to 10 mm long; seeds about 2 mm long.

COMMON NAME: White Prickly Poppy.

HABITAT: Waste ground.

RANGE: Native of the southwestern United States; occasionally adventive eastward.

ILLINOIS DISTRIBUTION: Occasional throughout the state; a collection from a sandy prairie in Morgan County may represent a native population.

This species is readily distinguished from A. *mexicana* by its white flowers, greater number of stamens, and variegated color of the leaves. Illinois specimens previously referred to as A. *intermedia* belong here.

Flowering time is from June to September.

3. *Macleaya* R. Br.–Plume Poppy

Herbaceous perennials with yellow sap; leaves cauline, opposite, lobed; flowers in terminal panicles; sepals 2, free, petallike; petals

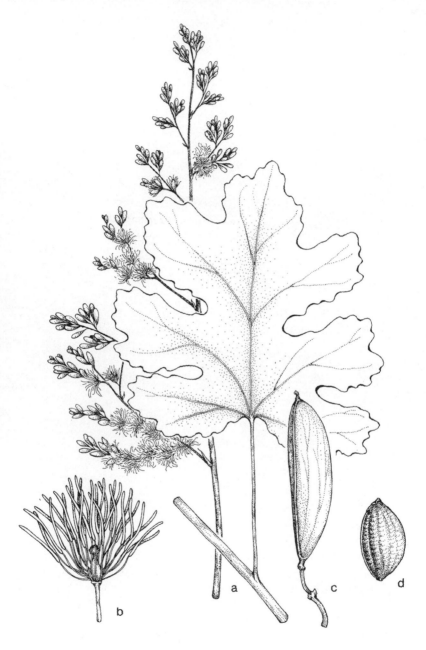

88. *Macleaya cordata* (Plume Poppy).   *a*. Leaf and inflorescence, X1.   *b*. Flower,
X3.   *c*. Fruit, X2½.   *d*. Seed, X5.

absent; stamens numerous, the anthers longitudinally dehiscent; pistil one, the ovary superior, the 2 stigmas spreading; capsule 2-valved, longitudinally dihiscent to the base, with 1–several seeds.

Only two species, both native to China and Japan, comprise the genus.

Only the following species has been found in Illinois.

1. **Macleaya cordata** (Willd.) R. Br. in App. Denh. & Clapp. Trav. 218. 1826. *Fig.* 88.

Bocconia cordata Willd. Sp. Pl. 2:841. 1800.

Robust perennial; stems erect, to 2.5 m tall, glabrous, more or less glaucous, with yellow sap; leaves alternate, 5- to 7-lobed, the lobes sinuate or dentate, to 30 cm long, cordate at the base, glabrous or pubescent beneath, glaucous beneath, on long, glabrous petioles; panicles terminal, to 30 cm long, composed of several flowers; sepals 2, free, creamy; petals none; stamens about 30; capsule ovoid, dehiscing all the way to the base, with 4–6 seeds.

COMMON NAME: Plume Poppy.

HABITAT: Waste ground.

RANGE: Native of China and Japan; occasionally adventive in the United States.

ILLINOIS DISTRIBUTION: Known only from Coles, Cook, and Henry counties.

The handsome plume poppy is rarely found outside of flower gardens.

The three adventive collections made in Illinois, in 1947, 1949, and 1977, represent accidental strays from cultivation. The Illinois specimens were flowering in July.

4. Papaver L.–Poppy

Annual, biennial, or perennial herbs, with milky sap; leaves cauline, alternate, lobed or dissected; flower showy, solitary, pedunculate; sepals 2 (–3). free; petals 4, free; stamens numerous, free; pistil one, the ovary superior, the stigmas united, 4- to 20-radiate; fruit a capsule dehiscing by terminal pores, with several seeds.

There are about fifty species of *Papaver* in the world, mostly all native of the Old World, with a few native to the western United States.

Several species of *Papaver* are grown as garden ornamentals because of their handsome flowers. These include *P. rhoeas* L., the corn poppy, and its many variants, known as Shirley poppies; *P. orientale* L., the giant scarlet oriental poppy; *P. glaucum* Boiss. & Haussk., the tulip poppy; *P. nudicaule* L., the Iceland poppy; and *P. somniferum* L., the opium poppy. This latter species is also the source of opium, obtained from the dried milky sap of the immature seed pod.

KEY TO THE SPECIES OF Papaver IN ILLINOIS

1. Stems glabrous; cauline leaves cordate-clasping; plants glaucous; capsule nearly spherical _____ 1. *P. somniferum*
1. Stems hirsute; cauline leaves not cordate-clasping; plants not glaucous; capsule longer than broad _____ 2
    2. Capsule narrowly obovoid _____ 2. *P. dubium*
    2. Capsule broadly obovoid _____ 3. *P. rhoeas*

1. Papaver somniferum L. Sp. Pl. 508. 1753. *Fig. 89.*

Rather stout annual, with white juice; stems erect, to 1.3 m tall, glabrous or somewhat pubescent, glaucous; leaves more or less oblong, dentate or pinnately lobed, glabrous or slightly pubescent, to about 25 cm long, the lower on short petioles, the upper clasping; flower solitary, on a long peduncle, white, pink, red, or purple, to 5 cm across; sepals 2; petals 4; stamens numerous; ovary superior; capsule globose, glabrous, up to 2.5 cm long.

COMMON NAME: Opium Poppy.

HABITAT: Waste ground.

RANGE: Native of Europe; occasionally escaped from cultivation in the eastern United States.

ILLINOIS DISTRIBUTION: Known as an adventive in Crawford County, where it was found "in original prairie along railroad," four miles northeast of Robinson, on June 11, 1932, by Pepoon and Barrett; also Christian, Coles, and Will counties.

It is the dried milky juice of the unripened capsule which is the source of opium. Two of the alkaloid constituents of opium are morphine and codeine.

Seeds of the opium poppy are used for putting on bread and rolls and also as a source of birdseed.

This species blooms from June to August.

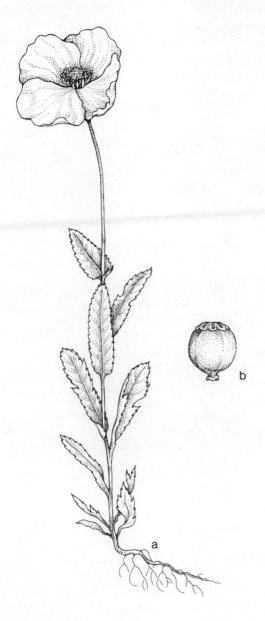

89. *Papaver somniferum* (Opium Poppy).   *a*. Habit, X½.   *b*. Capsule, X⅜.

2. **Papaver dubium** L. Sp. Pl. 1196. 1753. *Fig. 90*.

Slender annual, with white juice; stems erect, branched, to 60 cm tall, hirsute; leaves deeply pinnately lobed, hirsute, to about 15 cm long, the lower on long petioles, the upper sessile or nearly so; flower solitary on a long, pubescent peduncle, scarlet with usually a darkened center, to 5 cm across; sepals 2; petals 4; stamens numerous; ovary superior; capsule narrowly obovoid, glabrous, to 2 cm long, with 6–10 stigmatic rays across the disk at the top.

COMMON NAME: Smooth-fruited Poppy.

HABITAT: Waste ground.

RANGE: Native of Europe; occasionally escaped from cultivation in the eastern United States and the West Indies.

ILLINOIS DISTRIBUTION: Known as an adventive from Christian and Wabash counties.

This infrequently cultivated garden ornamental is even less often found as an adventive. The first Illinois collection seen was made in 1879.

This species is very similar in appearance to *P. rhoeas*, differing by its narrower capsule with fewer stigmatic rays across the disk at the top.

Papaver dubium flowers from May to August.

3. **Papaver rhoeas** L. Sp. Pl. 507. 1753. *Fig. 91*.

Slender annual, with white juice; stems erect, branched, to nearly 1 m tall, hispid; leaves pinnately lobed, hispid, to about 15 cm long, the lower on long petioles, the upper sessile or nearly so; flower solitary, on a long, hispid peduncle, scarlet with a darkened center, sometimes to nearly 10 cm across; sepals 2; petals 4; stamens numerous; ovary superior; capsule broadly obovoid, turbinate, glabrous, to 2 cm long, with 10 or more stigmatic rays across the disk at the top.

COMMON NAME: Corn Poppy.

HABITAT: Waste ground.

RANGE: Native of Europe; occasionally adventive in most areas of the United States.

ILLINOIS DISTRIBUTION: Scattered in various parts of Illinois.

The corn poppy is the most frequently encountered adventive poppy in Illinois.

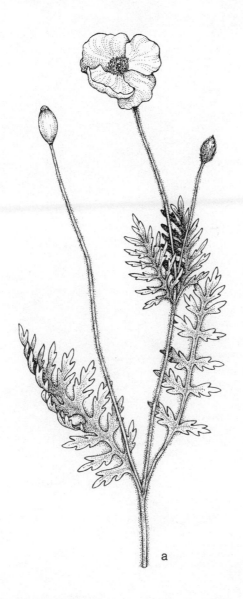

90. Papaver dubium (Smooth-fruited Poppy).   *a*. Habit, X½.

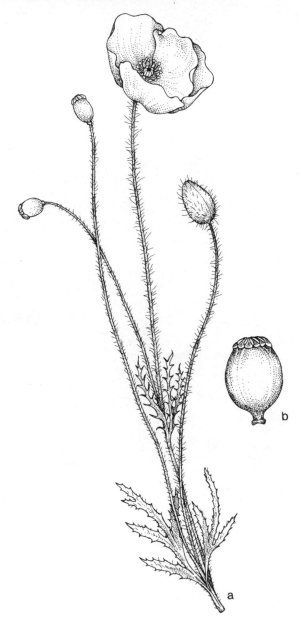

91. *Papaver rhoeas* (Corn Poppy). *a*. Habit, X½. *b*. Fruit, X1¼.

It is similar to *P. dubium*, differing by its broader capsule with ten or more stigmatic rays at the top. The corn poppy flowers from May to September.

5. *Stylophorum Nutt.*–Celandine Poppy

Perennial herbs, with yellow sap; leaves basal and also with one opposite pair on the stem, pinnately lobed; flowers 1–several, terminal; sepals 2, free; petals 4, free; stamens numerous, free; pistil one, the ovary superior, the stigma 2- to 4-lobed, radiate; fruit a capsule dehiscing all the way to the base, with several seeds.

Stylophorum is a genus of four handsome flowering plants occurring in eastern North America and Asia. The genus is most closely related to *Chelidonium*, which differs by its smaller flowers and narrower fruits.

Only the following species occurs in Illinois.

1. **Stylophorum diphyllum** (Michx.) Nutt. Gen. 2:7. 1818. *Fig.* 92.

Chelidonium diphyllum Michx. Fl. Bor. Am. 1:309. 1803.

Perennial herb with yellow sap, from a stout rootstock; stems erect, to 45 cm tall, sparsely hairy, more or less glaucous; leaves deeply pinnatifid, or the basal leaves with a pair of distinct leaflets, to 25 cm long, sparsely pubescent, glaucous on the lower surface; flowers 2–4, terminal, yellow, to 5 cm across; sepals 2, green, hirsute; petals 4, obovate to suborbicular; stamens numerous; capsule nodding, ovoid, 2- to 4-valved, hispid, to 2.5 cm long.

COMMON NAME: Celandine Poppy.

HABITAT: Moist, rich woods.

RANGE: Western Pennsylvania to Wisconsin, south to Missouri, Tennessee, and Virginia.

ILLINOIS DISTRIBUTION: Occasional in the southern one-fourth of the state; also Cook and Vermilion counties.

This handsome species is found in rich, usually beech-maple woods, associated with *Trillium gleasonii, Polemonium reptans, Sanguinaria canadensis,* and other early spring wildflowers.

It flowers from March to May. The petals fall away relatively quickly.

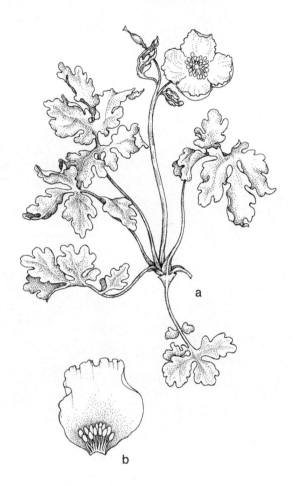

92. *Stylophorum diphyllum* (Celandine Poppy). *a*. Habit, X⅓. *b*. Petal, with some stamens, X1.

6. Chelidonium L.–Celandine

Biennial herb, with yellow sap; leaves basal and cauline, pinnately divided; flowers several, in terminal umbels; sepals 2, free; petals 4, free; stamens numerous, free; pistil one, the ovary superior, the stigma 2-lobed; fruit a capsule dehiscing from the base upward, with several seeds.

Only the following species comprises this genus.

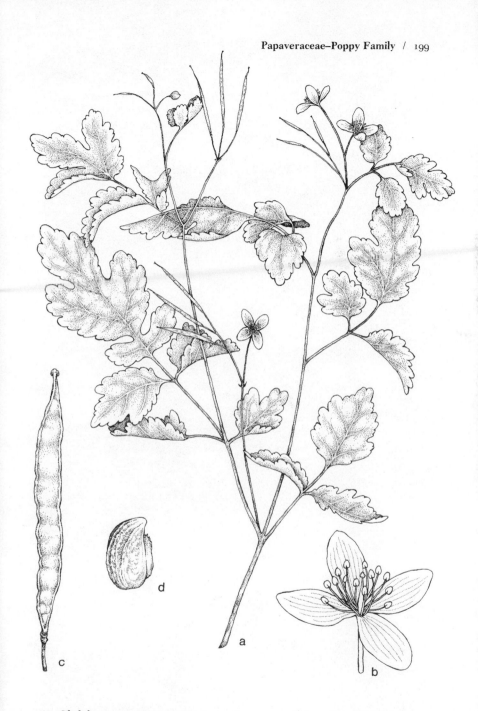

93. *Chelidonium majus* (Celandine). *a*. Upper part of plant, X⅜. *b*. Flower, X3.
c. Fruit, X4. *d*. Seed, X10.

1. **Chelidonium majus** L. Sp. Pl. 505. 1753. *Fig. 93*.

Biennial herb with yellow sap, from a stout rootstock; stems erect, branched, to 70 cm tall, brittle, sparsely pubescent; leaves pinnately divided, to 20 cm long, usually glabrous, glaucous on the lower surface; flowers to 1.8 cm across, yellow, several in a pedunculate umbel, the rays of the umbel to 1.5 cm long, pubescent; sepals 2, usually puberulent; petals 4, broadly elliptic, obtuse; stamens numerous; capsule erect, linear, 2-valved, glabrous, to 2.5 cm long, tipped with the persistent style and stigma.

COMMON NAME: Celandine.

HABITAT: Waste areas.

RANGE: Native of Europe; occasionally adventive in the United States.

ILLINOIS DISTRIBUTION: Scattered but not common in Illinois.

The celandine is a species at one time cultivated rather commonly as a garden ornamental. From these gardens it has occasionally escaped into waste ground.

This species flowers from May to August.

7. *Eschscholtzia Cham.*–California Poppy

Annual or perennial herbs, with watery juice; leaves ternately or pinnately divided; flower solitary, on long peduncles; sepals 2, coherent, hoodlike; petals 4, free, attached to the hollowed receptacle which surrounds the pistil; stamens numerous, free; pistil one, the ovary superior, the stigma 4- to 6-lobed; fruit a capsule dehiscing from the base upward, with several seeds. `

Eschscholtzia is a genus of handsome herbs native to western North America. Several of the species are cultivated as garden ornamentals.

Only the following species has been recorded from Illinois.

1. **Eschscholtzia californica** Cham. in Nees, Hors. Phys. Ber. 73. 1820. *Fig. 94*.

Perennial or sometimes annual herb; stems erect, often branched, to 65 cm tall, glabrous, glaucous; leaves ternately divided into linear or oblong segments, glabrous, glaucous, on long peduncles; flower solitary, to 6 cm across, yellow or orange; sepals 2, coherent into a hood which is pushed off by the expanding petals; petals 4,

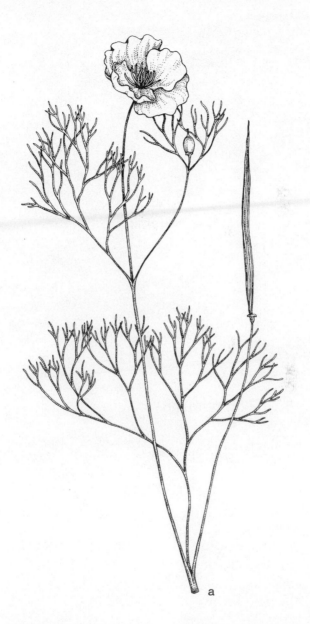

94. *Eschscholtzia californica* (California Poppy).   *a*. Habit, in flower and in fruit, X⅜.

free, inserted on the rim of the receptacle; stamens numerous; capsule erect, linear, 2-valved, glabrous, strongly ribbed, to 8 cm long.

COMMON NAME: California Poppy.
HABITAT: Waste areas.
RANGE: California; occasionally adventive eastward.
ILLINOIS DISTRIBUTION: This adventive is known from two Illinois collections, both made during the latter part of the nineteenth century.

The California poppy is a handsome species often cultivated as a garden ornamental.

The distinguishing feature of the genus is the hood-like connivent sepals which are pushed off as the petals expand. This species flowers from June to September.

8. Dicentra Bernh.–Bleeding-heart

Perennial herbs, usually from small bulbs or tubers; leaves often only basal, ternately compound with much divided leaves; flowers nodding, in racemes, on bibracteate pedicels; sepals 2, free; petals 4, connivent in 2 pairs, the exterior pair spurred at the base and spreading at the apex, the inner pair clawed and crested or winged on the back; stamens 6, in two sets of three each, the filaments of each set slightly united; pistil one, the ovary superior, the stigma 2-crested or 2-horned; fruit a capsule dehiscing all the way to the base, with several seeds.

This genus of peculiarly shaped flowers is composed of about fifteen species native to North America and Asia. Many of the species are grown as garden ornamentals, including *D. spectabilis* Lem., the rosy-red bleeding-heart.

KEY TO THE SPECIES OF Dicentra IN ILLINOIS

1. Spurs of corolla spreading, subacute; flowers without an odor; plants from granular, white tubers _____ 1. *D. cucullaria*
1. Spurs of corolla not spreading, rounded; flowers with a sweet odor; plants from yellow, cornlike tubers _____ 2. *D. canadensis*

1. **Dicentra cucullaria** (L.) Bernh. Linnaea 8:457. 1833. *Fig. 95.*

Fumaria cucullaria L. Sp. Pl. 699. 1753.
Dielytra cucullaria (L.) Torr. & Gray, Fl. N. Am. 1:66. 1838.
Bicuculla cucullaria (L.) Millsp. Bull. W. Va. Agr. Exp. Sta. 2:327. 1892.

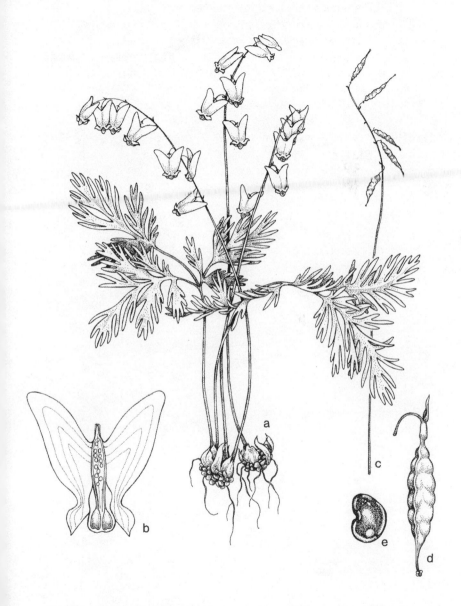

95. *Dicentra cucullaria* (Dutchman's-breeches). *a*. Habit, in flower, X½. *b*. Section through flower, X3. *c*. Fruiting stalk, X½. *d*. Fruit, X2½. *e*. Seed, X5.

Perennial herb from granular, white tubers; leaves all basal, ternately compound, the ultimate divisions linear, glabrous, pale on the lower surface; scape to 25 cm tall, slender, bearing up to 10 flowers; flowers white or faintly pink, yellow at the summit, not aromatic, nodding, to 1.8 cm long, on very slender pedicels; spurs subacute, widely spreading; crest of inner petals minute; capsule broadly fusiform, glabrous, to 2 cm long, tipped by the persistent style and stigma.

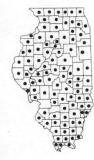

COMMON NAME: Dutchman's-breeches.

HABITAT: Rich woods.

RANGE: Nova Scotia to North Dakota, south to Kansas, Alabama, and Georgia; Washington; Oregon.

ILLINOIS DISTRIBUTION: Common throughout the state; in almost every county.

The delightful little odd flowers, shaped like a pair of inflated breeches, are familiar to all enthusiasts of nature. The nodding blossoms, which are without an odor, open in March, April, and May, a full week before the flowers of *D. canadensis* open.

The white, granular tubers are said to be poisonous to cattle.

The first reference to this plant in Illinois was by Mead, who referred to it in 1846 as *Dielytra cucullaria* (L.) Torr. & Gray.

2. Dicentra canadensis (Goldie) Walp. Rep. 1:118. 1842. *Fig. 96.*

Corydalis canadensis Goldie, Edinb. Phil. Journ. 6:329. 1822.

Bicuculla canadensis (Goldie) Millsp. Bull. W. Va. Agr. Exp. Sta. 2:327. 1892.

Perennial herb from yellow, cornlike tubers; leaves all basal, ternately compound, the ultimate divisions linear, glabrous, pale on the lower surface; scape to 25 cm tall, slender, bearing up to 8 flowers; flowers greenish-white, often tinged with purple, sweetly scented, nodding, to 2 cm long, on very slender pedicels; spurs rounded, not spreading; crest of inner petals conspicuously projecting; capsule broadly fusiform, glabrous, to 2.5 cm long, tipped by the persistent style and stigma.

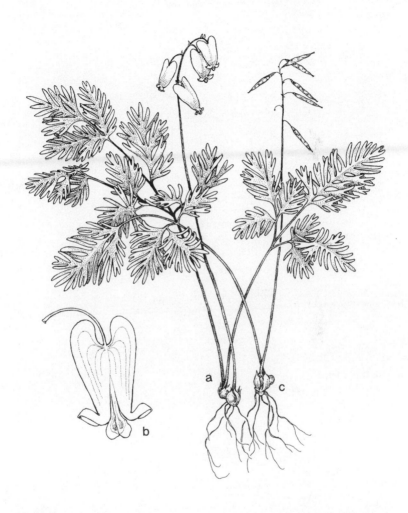

96. Dicentra canadensis (Squirrel-corn). *a*. Habit, in flower, X½. *b*. Flower, X2. *c*. Fruiting stalk, X½.

COMMON NAME: Squirrel-corn.

HABITAT: Rich woods.

RANGE: Quebec to Minnesota, south to Missouri and North Carolina.

ILLINOIS DISTRIBUTION: Occasional and scattered throughout Illinois.

This species is much less common than *D. cucullaria*, although it usually grows in similar habitats.

Although it is difficult if not impossible to distinguish *D. canadensis* from *D. cucullaria* on the characters of the leaves and fruits, there are several other reliable differences. In *D. canadensis*, the flowers are sweetly scented, like a hyacinth, the spurs of the outer petals are rounded and not divergent, the crest on the inner petals projects conspicuously, and the underground tubers look like kernels of corn.

Squirrel-corn, on the average, begins to bloom about one week later than Dutchman's-breeches, from early April to May.

The cornlike tubers are reputedly poisonous to livestock.

9. *Adlumia Raf.*–Climbing Fumitory

Annual or biennial vine, climbing by slender petioles; leaves tripinnately compound; flowers numerous in axillary, pendent panicles; sepals 2, free; petals 4, united into a persistent, spongy corolla; stamens 6, united at the base, but splitting part way up in two clusters of three stamens each; pistil one, the ovary superior, the stigma 2-crested; fruit a capsule dehiscing all the way to the base, enclosed by the persistent corolla, with few seeds.

Only the following species comprises the genus.

1. **Adlumia fungosa** (Ait.) Greene in BSP. Prel. Cat. N.Y. 3. 1888. *Fig.* 97.

Fumaria fungosa Ait. Hort. Kew. 3:1. 1789.

Adlumia cirrhosa Raf. Mod. Rep. 5:352. 1808.

Delicate, biennial vine climbing over bushes; leaves tripinnately compound, the leaflets ovate, rounded or cuneate at the base, entire or more commonly 3-lobed, glabrous, pale beneath; flowers numerous, in axillary, pendent panicles, greenish-purple, to 1.5 cm long; corolla 4-lobed at the apex, narrowly ovate; filaments adherent to the corolla; capsule linear, 2-valved, enclosed by the persistent corolla, with few seeds.

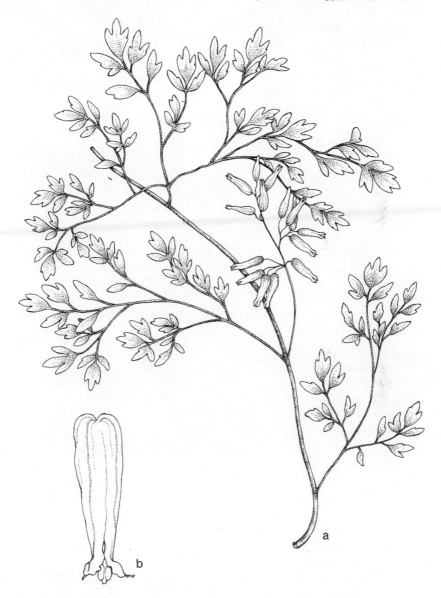

97. *Adlumia fungosa* (Climbing Fumitory).   *a*. Habit, X½.   *b*. Flower, X3.

COMMON NAME: Climbing Fumitory; Allegheny Vine.

HABITAT: In woods, after a fire (according to Swink 1974).

RANGE: Quebec to Minnesota, south to Kentucky and North Carolina; adventive elsewhere in the eastern United States.

ILLINOIS DISTRIBUTION: Known at least from Kankakee and Ogle counties.

This delicate vine, which climbs by means of its petioles, is sometimes planted as a garden ornamental, but rarely escapes into disturbed areas. A specimen collected by E. J. Hill from Kankakee County has been seen, and a report by Patterson (1876) of a Bebb collection from Ogle County is probably accurate. Two other specimens reputedly from Illinois have been seen. One, collected by F. Brendel and reported by him in 1887 (as *Adlumia cirrhosa*), merely has "North Illinois" as its locality. The other, collected by E. R. Boardman, has "Southern Illinois, rocky hills, July" as its locality.

The climbing fumitory flowers from June to September.

10. *Corydalis Medic.*–Corydalis

Biennial or perennial herbs, rarely climbing; leaves pinnately decompound, basal and cauline; flowers in racemes, terminal or opposite the petioles; sepals 2, free; petals 4, connivent, one of the outer ones spurred at the base, one of the inner ones keeled; stamens 6, in 2 clusters of 3 each; pistil one, the ovary superior; fruit a capsule dehiscing all the way to the base, with several seeds.

Corydalis is a genus of more than one hundred species native to the north temperate regions of the world and to South Africa. It differs from *Dicentra* and *Adlumia* by its one-spurred corolla and from *Fumaria* by its elongated fruit.

The genus was monographed by Ownbey in 1947.

KEY TO THE SPECIES OF Corydalis IN ILLINOIS

1. Flowers pink, with yellow tips _____ 1. *C. sempervirens*
1. Flowers wholly yellow _____ 2
   2. Outer petals with winged crest down the back _____ 3
   2. Outer petals not crested down the back _____ 5
3. Winged crest of outer petals with 3–4 teeth; stalk of capsule 1 cm long or longer; seeds rugulose _____ 2. *C. flavula*
3. Winged crest of outer petals entire; stalk of capsule up to 5 mm long;

seeds smooth _____ 4

4. Capsules scarcely torulose, up to 15 mm long; uppermost raceme not overtopping subtending leaves _____ 3. *C. micrantha*

4. Capsules strongly torulose, 15–25 mm long; uppermost raceme overtopping subtending leaves _____ 4. *C. halei*

5. Spur ⅓–⅖ the length of the corolla; sepals up to 1 mm long; seeds 1.3–1.7 mm long _____ 7. *C. campestris*

5. Spur less than one-third the length of the corolla; sepals 1.5–2.0 mm long; seeds 2.0–2.5 mm long _____ 6

6. Capsules pendulous or widely spreading; margins of seeds rounded. _____ 5. *C. aurea*

6. Capsules ascending; margins of seeds sharp _____ 6. *C. montana*

1. **Corydalis sempervirens** (L.) Pers. Syn. 2:269. 1807. *Fig. 98.*

Fumaria sempervirens L. Sp. Pl. 2:700. 1753.

Corydalis glauca Pursh, Fl. Am. Sept. 463. 1814.

Annual or biennial herb; stem erect or ascending, branched, to 85 cm tall, glabrous, glaucous; leaves pinnately decompound, glabrous, glaucous, the ultimate segments more or less obovate, cuneate, entire or with mucronulate-tipped lobes, the lower leaves to 10 cm long and short-petiolate, the upper smaller and sessile or nearly so; flowers numerous in panicles, pink with a yellow tip, to 1.8 cm long, on slender pedicels; spur rounded, 1.5–2.5 mm long; capsules erect, linear, torulose, to 5 cm long, to 2 mm broad; seeds several, orbicular, shiny, reticulate.

COMMON NAME: Pink Corydalis.

HABITAT: Rocky woods; sandy soil.

RANGE: Newfoundland to Alaska, south to British Columbia, Montana, Minnesota, northern Illinois, and northern Georgia.

ILLINOIS DISTRIBUTION: Rare in the northern one-fourth of the state; known from Cook, LaSalle, and Ogle counties.

The pink corydalis is one of the handsomest as well as one of the rarest wild flowers in Illinois. The pink flower with a yellow tip, and the heavily glaucous foliage, are extremely beautiful.

Although Brendel records this species (as *C. glauca* Pursh) from Illinois in 1887, the earliest Illinois collection I have seen was made by H. C. Cowles on June 10, 1896, from Cook County. Since that

98. *Corydalis sempervirens* (Pink Corydalis). *a*. Upper part of plant, X½. *b*. Flower, X3. *c*. Fruit, X2. *d*. Seed, X7½.

time it has been found at Starved Rock State Park in LaSalle County in 1938, 1939, and 1941, and more recently in Ogle County in 1944 and 1946. There is a possibility that it is now extinct in Illinois.

The flowers bloom from May to mid-August.

2. **Corydalis flavula** (Raf.) DC. Prodr. 1:129. 1824. *Fig. 99.*

Fumaria flavula Raf. Journ. Bot. Desv. 1:224. 1808.

Annual herb; stem spreading to ascending, much branched, to 45 cm long, glabrous; leaves pinnately decompound, glabrous, the ultimate segments linear to oblong, usually cuneate, lobed, the lower leaves short-petiolate, the upper sessile or nearly so; flowers several in racemes, pale yellow, to 8 mm long, on slender pedicels; outer petals longer than the inner, the spur rounded, to 2 mm long, the crest of the outer petals winged and bearing 3–4 teeth; capsules ascending to drooping, linear-cylindric, torulose, to 2 cm long, on slender pedicels 1 cm long or longer; seeds several, orbicular, sharp-margined, rugulose.

COMMON NAME: Pale Corydalis.

HABITAT: Rich woods, often along streams.

RANGE: Connecticut to southern Ontario and Minnesota, south to Kansas, Louisiana, and Virginia.

ILLINOIS DISTRIBUTION: Common in the southern tip of the state, becoming rarer northward and possibly even extinct in the northeastern counties.

Pale corydalis differs from the similar-appearing *C. micrantha* and *C. halei* by its toothed crest of the outer petals, its long fruiting pedicels, and its rugulose seeds.

In low bottomland woods in the southernmost counties of Illinois, this species may form huge, dense colonies.

The flowers appear in April and May.

3. **Corydalis micrantha** (Engelm.) Gray, Bot. Gaz. 11:189. 1886. *Fig. 100.*

Corydalis aurea Willd. var. *micrantha* Engelm. ex Gray, Man. Bot. ed. 5, 62. 1867.

Capnoides micranthum (Engelm.) Britt. Mem. Torrey Club 5:166. 1894.

Corydalis micrantha (Engelm.) Gray ssp. *micrantha* Ownbey, Ann. Mo. Bot. Gard. 34:219. 1947.

Annual herb; stem spreading to ascending, much-branched, to 30

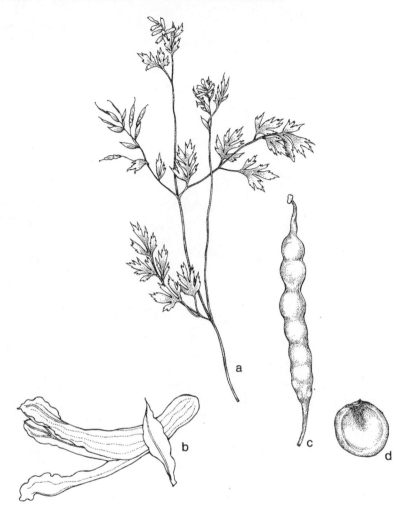

99. *Corydalis flavula* (Pale Corydalis). *a*. Upper part of plant, X⅓. *b*. Flower, X4. *c*. Fruit, X2½. *d*. Seed, X7½.

cm long, glabrous; leaves pinnately decompound, glabrous, the ultimate segments linear to oblong, usually cuneate, lobed, the lower leaves short-petiolate, the upper sessile or nearly so; flowers several in racemes, the uppermost raceme not overtopping the subtending leaves, pale yellow, the earliest flowers to 13 mm long, with a rounded spur to 1 mm long, on slender pedicels, the later flowers

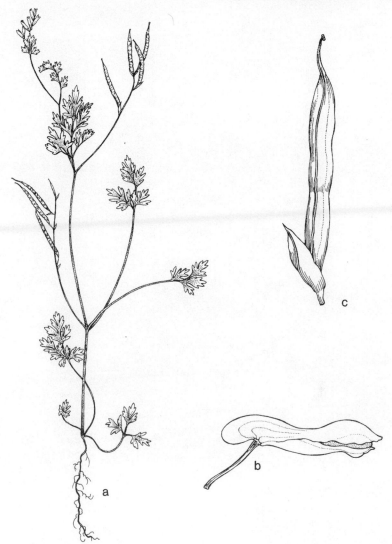

100. *Corydalis micrantha* (Slender Corydalis). *a*. Habit, X½. *b*. Flower, X3. *c*. Fruit, X5.

smaller and cleistogamous, spurless; outer petals longer than the inner, the crest winged and without teeth; capsules ascending to erect, linear-cylindric, slightly torulose, to 1.5 cm long, on slender pedicels up to 5 mm long; seeds several, orbicular, round-margined, smooth.

COMMON NAME: Slender Corydalis.

HABITAT: Rocky woods; sandy soil.

RANGE: Wisconsin to South Dakota, south to Texas and Tennessee.

ILLINOIS DISTRIBUTION: Occasional or rare but scattered throughout the state.

The slender corydalis is often confused with *C. flavula*, but differs by the toothless crest of the outer petals and by the smooth seeds. The even more similar but rarer *C. halei* differs by its longer, more torulose capsules and its elongated uppermost raceme.

Corydalis micrantha generally grows in less moist woods than does *C. flavula*. It flowers from May to July.

Early workers in Illinois, such as Lapham (1857), Patterson (1876), and Brendel (1887) erroneously reported this species as *C. aurea* Willd.

4. **Corydalis halei** (Small) Fern. & Schub. Rhodora 48:207. 1946. *Fig. 101*.

Corydalis aurea Willd. var. *australis* Chapm. Fl. So. U.S. Suppl. 1:604. 1883.

Capnoides halei Small, Bull. Torrey Club 25:137. 1898.

Corydalis micrantha (Engelm.) Gray ssp. *australis* (Chapm.) Ownbey, Ann. Mo. Bot. Gard. 34:222. 1947.

Corydalis micrantha (Engelm.) Gray var. *australis* (Chapm.) Shinners, Field & Lab. 18:42. 1950.

Annual herb; stem spreading to ascending, much-branched, to 40 cm long, glabrous; leaves pinnately decompound, glabrous, silvery-green, the ultimate segments linear to oblong, usually cuneate, lobed, the lower leaves short-petiolate, the upper sessile or nearly so; flowers several in racemes, the uppermost raceme to 20 cm long, overtopping the subtending leaves, pale yellow, the earliest flowers to 13 mm long, with a rounded spur to 1 mm long, on slender pedicels, the later flowers smaller and cleistogamous, spurless; outer petals longer than the inner, the crest winged and without teeth; capsules ascending to erect, linear-cylindric, strongly torulose, to 2.5 cm long, on slender pedicels up to 5 mm long; seeds several, orbicular, round-margined, smooth.

101. *Corydalis halei* (Slender Corydalis). *a*. Habit, X½. *b*. Flower, X3. *c*. Fruit, X2. *d*. Seed, X7½.

COMMON NAME: Slender Corydalis.

HABITAT: Moist soil at base of limestone cliff.

RANGE: Virginia to southeastern Missouri, south to Texas and Florida.

ILLINOIS DISTRIBUTION: Known only from Monroe Co.: one mile south of Fults, April 23, 1965, *J. Ozment s.n.*

This species is closely related to *C. micrantha* and is considered by Ownbey (1947) and others as a subspecies of it and by Shinners (1950) as a variety of it.

In addition to the distinguishing characters given in the key, *C. halei* usually has silvery-green leaves.

This is a species of the southeastern United States, particularly along the coastal plain.

The flowers appear in April and May.

5. Corydalis aurea Willd. Enum. Hort. Berol. 2:7 40. 1809. *Fig. 102.*

Capnoides aureum (Willd.) Kuntze, Rev. Gen. Pl. 14. 1891.

Annual or biennial herb; stem spreading to ascending, much-branched, to 50 cm long, glabrous, glaucous; leaves pinnately decompound, glabrous, glaucous, the ultimate segments oblong to obovate, usually cuneate, lobed, the lower leaves petiolate, the upper sessile or nearly so; flowers several in racemes, golden-yellow, to 2 cm long, on short pedicels; sepals 1.5–2.0 mm long; outer petals not crested down the back; spur rounded, slightly curved, less than one-third as long as the corolla; capsules spreading or drooping, on short, slender pedicels, linear, torulose, to 2.5 cm long; seeds several, orbicular, round-margined, smooth, 2.0–2.5 mm long.

COMMON NAME: Golden Corydalis.

HABITAT: Rocky woods.

RANGE: Quebec to Alaska, south to California, New Mexico, Texas, Missouri, and West Virginia.

ILLINOIS DISTRIBUTION: Known only from Champaign, Cook, Ford, LaSalle, Mason, Ogle, and Winnebago counties.

This handsome species, with golden-yellow flowers, is an inhabitant of rocky woods where it is rare in Illinois but known from almost half the counties of Missouri.

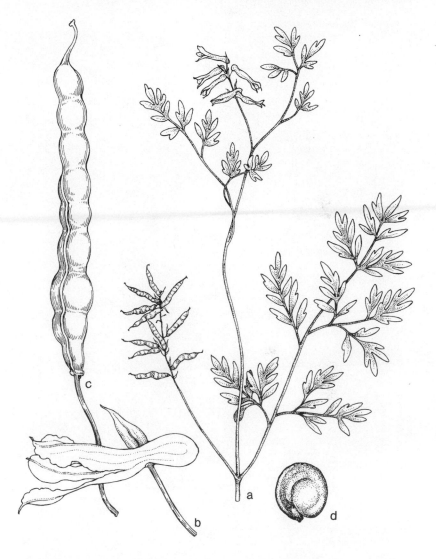

102. *Corydalis aurea* (Golden Corydalis).   *a*. Upper part of plant, X½.   *b*. Flower, X3.   *c*. Fruit, X4.   *d*. Seed, X7½.

The crestless outer petals relate this species very closely to *C. montana*, but this latter species differs by its longer spur, shorter sepals, and smaller seeds.

The golden corydalis blooms from May to August.

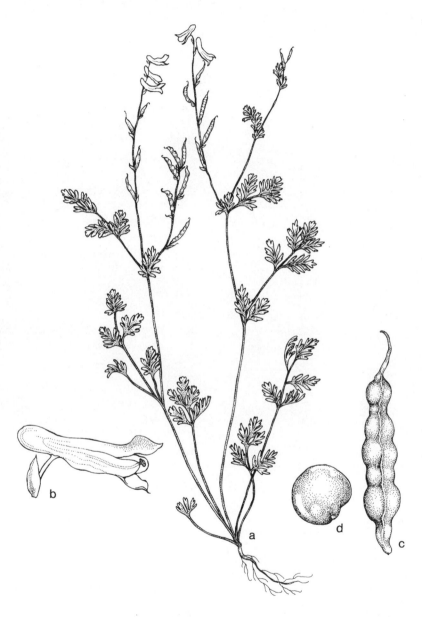

103. *Corydalis montana* (Corydalis).   *a*. Habit, X½.   *b*. Flower, X2½.   *c*. Fruit, X2½.   *d*. Seed, X10.

6. **Corydalis montana** Engelm. ex Gray, Mem. Am. Acad. 4:6. 1849. *Fig. 103.*

Corydalis aurea Willd. var. *occidentalis* Engelm. ex Gray, Man. ed. 5, 62. 1867.

Capnoides montanum (Engelm.) Britt. Mem. Torrey Club 5:166. 1894.

Annual or biennial herb; stems spreading to ascending, much-branched, to 50 cm long, glabrous, more or less glaucous; leaves pinnately decompound, glabrous, glaucous, the ultimate segments oblong to obovate, usually cuneate, lobed, the lower leaves petiolate, the upper sessile or nearly so; flowers several in racemes, golden-yellow, to 2 cm long, on very short pedicels; sepals up to 1 mm long; outer petals not crested down the back; spur rounded, slightly curved, one-third to two-fifths as long as the corolla; capsules ascending, on short, slender pedicels, linear, torulose, to 2.5 cm long; seeds several, orbicular, sharp-margined, smooth, 1.3–1.7 mm long.

COMMON NAME: Corydalis.

HABITAT: Rocky woods.

RANGE: Northern Illinois to Montana, south to Texas and Missouri; Mexico.

ILLINOIS DISTRIBUTION: Occasional in the northern one-third of the state, rare elsewhere.

Several authors consider *C. montana* to be merely a variety of *C. aurea* and, indeed, these two taxa are very much alike. However, I believe that the differences outlined in the key to species are sufficient enough to merit specific status for *C. montana*.

Ownbey (1947), who treats this taxon as a variety, records its range considerably west of Illinois, but I have seen specimens from Illinois which I identify as *C. montana*.

The flowers open in April and May.

7. **Corydalis campestris** (Britt.) Buchholz & Palmer, Trans. Acad. Sci. St. L. 25:115. 1926. *Fig. 104.*

Capnoides campestre Britt. Man. ed. 2, 1065. 1905.

Annual or biennial herb; stems more or less erect, branched, to 50 cm tall, glabrous, more or less glaucous; leaves pinnately decompound, glabrous, glaucous, the ultimate segments oblong to ob-

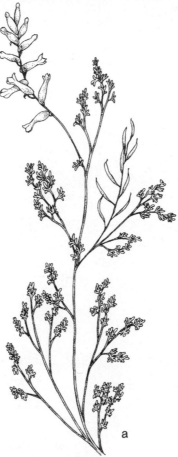

104. Corydalis campestris (Plains Corydalis). *a.* Upper part of plant in flower and fruit, X½.

ate, usually cuneate, lobed, the lower leaves petiolate, the upper sessile or nearly so; flowers several in racemes, bright yellow, to 2 cm long, on very short pedicels; sepals up to 2 mm long; outer petals not crested down the back; spur rounded, nearly straight, one-third to one-half as long as the corolla; capsules strongly ascending, on short pedicels, thickly linear, more or less torulose, to 2.5 cm long; seeds several, orbicular, sharp-margined, with conspicuous rings of reticulations, 1.8–2.7 mm long.

COMMON NAME: Plains Corydalis.

HABITAT: Prairies.

RANGE: Illinois to Nebraska, south to Texas and Arkansas.

ILLINOIS DISTRIBUTION: Known only from Morgan County, where it was collected in 1928.

This primarily prairie species is closely related to *C. aurea* and *C. montana*, but differs from both by its seeds which are distinctly reticulate in concentric rings. In addition, the capsules, on the average, are thicker in *C. campestris* than in either of the other two.

There is some difference of opinion as to the status of *C. campestris*. Although I recognize it as a distinct species, there are others who would equate it with *C. montana*. The characters of the seed, however, would seem to preclude this latter disposition.

The flowers of the plains corydalis open in May and June.

11. Fumaria L.–Fumitory

Annual herbs; leaves alternate, pinnately decompound; flowers numerous, in racemes or spikes; sepals 2, free; petals 4, connivent, one of the outer ones spurred at the base, the inner ones keeled or crested; stamens 6, in two clusters of three each; pistil one, the ovary superior; fruit small, globose, 1-seeded, indehiscent.

Fumaria is composed of about forty species, all native to the Old World.

The genus is similar to *Corydalis* except for the remarkable globose, indehiscent, one-seeded fruits.

Only the following species has been found in Illinois.

1. Fumaria officinalis L. Sp. Pl. 700. 1753. *Fig. 105.*

Annual herb from fibrous roots; stems spreading or ascending, much-branched, to 1 m long, glabrous; leaves pinnately decompound, glabrous, petiolate, the ultimate segments linear to oblong, cuneate, entire or lobed; flowers numerous, to 7 mm long, pale purple tipped with crimson, pedicellate, in both terminal and axillary racemes, the racemes to 7 cm long; sepals ovate-lanceolate, acute, serrulate, about 2 mm long; spur rounded, slightly curved, about 1 mm long; fruit globose, slightly retuse at summit, 1.5–2.0 mm in diameter, glabrous.

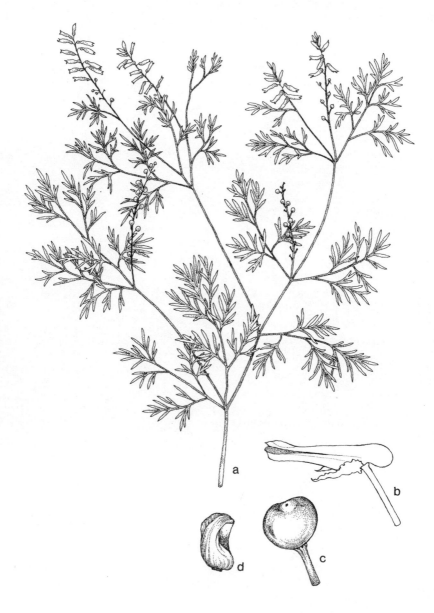

105. *Fumaria officinalis* (Fumitory). *a*. Upper part of plant, X½. *b*. Flower, X5. *c*. Fruit, X7½. *d*. Seed, X12½.

COMMON NAME: Fumitory.

HABITAT: Waste ground, often in the shade.

RANGE: Native of Europe; occasionally adventive in North America.

ILLINOIS DISTRIBUTION: Occasional and scattered in northern Illinois, rare elsewhere.

Fumitory is sometimes grown as a garden ornamental in Illinois. Although it sometimes escapes, it is not a persistent adventive.

The flowers, which are small but interestingly colored, bloom from May to August.

Order Nymphaeales

I have departed from Thorne's classification (1968) by elevating his subfamilies of Nymphaeaceae to family status. Thus, subfamily Nymphaeoideae becomes the Nymphaeaceae, Nelumbonoideae becomes the Nelumbonaceae, and Cabomboideae becomes the Cabombaceae. In addition, the Ceratophyllaceae is included in this order.

NYMPHAEACEAE–WATER LILY FAMILY

Aquatic plants from stout rhizomes; leaves submersed, floating, or emersed, usually cordate, long-petiolate; flowers solitary on long peduncles; sepals 4–6, free, green or reddish-tinged; petals numerous, free, often brightly colored; stamens numerous, often inserted on the ovary, usually with broad, flat filaments; ovary compound, with several locules and several ovules; stigmas radiating; fruits leathery, at length breaking irregularly.

As constituted here, the Nymphaeaceae are composed of six genera and less than 100 species.

Botanists are not in accord as to the makeup of the family. Frequently the genera *Nelumbo*, *Cabomba*, and *Brasenia* are included in this family. Recent experimental evidence seems to indicate that these three genera should be segregated from *Nuphar* and *Nymphaea*. This latter view is followed in The Illustrated Flora.

KEY TO THE GENERA OF Nymphaeaceae IN ILLINOIS

1. Flowers yellow; sepals 5–6; leaves oval _____ 1. *Nuphar*
1. Flowers white or pinkish; sepals 4; leaves orbicular __ 2. *Nymphaea*

1. *Nuphar Smith*–Yellow Water Lily

Aquatic perennials from stout rhizomes; leaves submersed, floating, or emersed, cordate, long-petiolate; flowers solitary, on long, naked peduncles; sepals 5–6, free, strongly concave, arranged to form a cup; petals numerous, free, yellow, shorter than the sepals; stamens numerous, with flat filaments and broad, flat connectives; ovary compound, with many locules and many ovules, the stle thick, the stigmas broad, 6- to 24-rayed; fruit leathery, at length breaking irregularly.

Beal (1956) has revised the North American and European species, and his work is followed here.

Only the following species occurs in Illinois.

1. **Nuphar luteum** (L.) Sibth. & Small, Fl. Graec. Prodr. 1:361. 1808.

Nymphaea lutea L. Sp. Pl. 510. 1753.

Aquatic perennial with thick, horizontal, cylindrical rootstocks; floating and emersed leaves usually oblong to ovate, cordate at the base, the basal lobes sometimes overlapping, glabrous above, glabrous or puberulent below, to 30 cm long, to 25 cm across, on terete or flattened, often winged petioles; submerged leaves thin, orbicular, cordate at the base; flowers yellow, to 5 (–7) cm across; sepals usually 6, sometimes tinged with red; petals varying from thin to thick, oblong, truncate, yellow; stamens numerous, in several rows; stigmatic disk crenate, yellow and sometimes tinged with red, with 5–26 rays; fruit ovoid, sometimes constricted into a short neck at the top, to 4 cm long; seeds ovoid, to 6 mm long.

Two subspecies may be recognized in Illinois.

1. Petiole conspicuously flattened and winged on the upper surface; sepals usually red-tinged within _____ 1a. *N. luteum* ssp. *variegatum*
1. Petiole terete to more or less flattened; sepals rarely red-tinged within _____ 1b. *N. luteum* ssp. *macrophyllum*

1a. **Nuphar luteum** (L.) Sibth. & Small ssp. **variegatum** (Engelm. ex Clinton) Beal, Journ. Elisha Mitch. Soc. 72:330. 1956. *Fig. 106.*

Nymphaea advena Ait. Hort. Kew. 2:226–27. 1788, in part.

Nuphar advena (Ait.) Ait. f. Hort. Kew. ed. 2, 3:295. 1811, in part.

Nuphar variegatum Engelm. ex Clinton, 19th Ann. Rep. Reg. State Univ. N.Y. 73. 1866.

Nuphar advena (Ait.) Ait. f. var. *variegatum* (Engelm.) Gray, Man., ed. 5, 57. 1868.

Petiole conspicuously flattened and winged on the upper surface; sepals usually red-tinged within; stigmatic rays often extending to the margin of the disk.

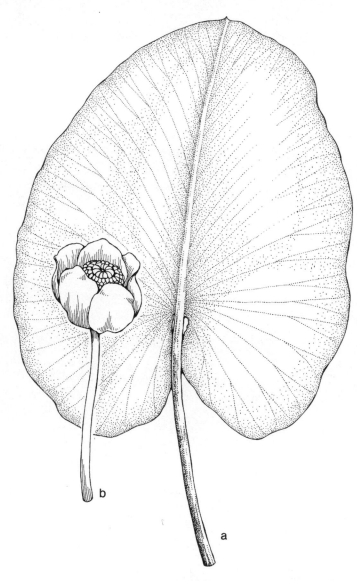

106. Nuphar luteum ssp. *variegatum* (Pond Lily).   *a*. Leaf, X½.   *b*. Flower, X½.

COMMON NAME: Pond Lily.
HABITAT: Ponds; shallow water of lakes and streams.
RANGE: Newfoundland to Yukon, south to Montana, Nebraska, Iowa, northern Illinois, northern Ohio, and Delaware.
ILLINOIS DISTRIBUTION: Known only from Cook, Grundy, Lake, and McHenry counties.

This plant is often considered a distinct species, usually referred to as *N. variegatum*. Beal's work (1956) shows that it is best treated as a subspecies of *Nuphar luteum*. The typical ssp. *luteum*, a native of Europe, is narrowly constricted below the stigmatic disk, and often has only five sepals.

The flowers are produced from June to September.

1b. Nuphar luteum (L.) Sibth. & Small ssp. **macrophyllum** (Small) Beal, Journ. Elisha Mitch. Sci. Soc. 72:332. 1956. *Fig. 107*.

Nymphaea advena Ait. Hort. Kew. 2:226–27. 1788, in part.
Nuphar advena (Ait.) Ait. f. Hort. Kew, ed. 2, 3:295. 1811, in part.
Nymphaea macrophylla Small, Bull. Torrey Club 25:465–66. 1898,
Nymphaea advena Ait. ssp. *macrophyllum* (Small) Miller & Standl. Contrib. U.S. Nat. Herb. 16:89–90. 1912.
Nuphar advena (Ait.) Ait. f. var. *brevifolium* Standl. Rhodora 31:37. 1929.

Petiole terete to more or less flattened; sepals rarely red-tinged; stigmatic rays usually extending to the margin of the disk.

COMMON NAME: Yellow Pond Lily.
HABITAT: Ponds; shallow water of lakes and streams.
RANGE: Southern Maine to southern Wisconsin, south to Texas and Florida; northeastern Mexico; Cuba.
ILLINOIS DISTRIBUTION: Scattered throughout the state.

This plant has commonly been known as *N. advena*. I am following Beal (1956) who considers this plant to be a subspecies of the European *N. luteum*.

A variant with smaller leaves, called var. *brevifolium* by Standley, has its type locality in Richland County where it was discovered by Robert Ridgway in

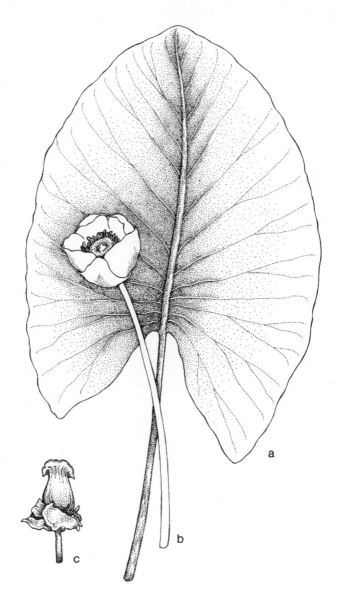

107., *Nuphar luteum* ssp. *macrophyllum* (Yellow Pond Lily).   *a*. Leaf, X½.   *b*. Flower, X½.   *c*. Flower, with perianth removed, X1.

1928. I do not believe it to be worthy of recognition.
The flowers appear from June to September.

2. *Nymphaea* L.–*Water Lily*

Aquatic perennials with creeping rhizomes; leaves large, alternate from along the rhizome, long-petiolate, with a deep basal sinus; flowers bisexual, showy, white or pink (in Illinois), on stout peduncles; sepals 4, free, green; petals numerous, free, gradually passing into the stamens; stamens numerous, free, borne on the ovary; pistil 1, the ovary superior, 12- to 35-locular, with many ovules, the stigma sessile, several-rayed; fruit a depressed-globose berry, enveloped by an aril, maturing under water.

Nymphaea is comprised of about forty wild species in the temperate and tropical regions of the world, in addition to a number of horticultural forms.

KEY TO THE SPECIES OF Nymphaea IN ILLINOIS

1. Petals subacute at tip; flowers fragrant; seeds 2 mm long _____
   _____ 1. *N. odorata*
1. Petals rounded at tip; flowers not fragrant; seeds 3–4 mm long _____
   _____ 2. *N. tuberosa*

1. Nymphaea odorata Ait. Hort. Kew. 2:227. 1789. *Fig. 108.*

Castalia odorata (Ait.) Woodville & Wood in Rees, Cycl. 6, no. 1. 1806.

Nymphaea odorata var. *gigantea* Tricker, Water Garden 186. 1897.

Aquatic perennial from a stout, forking rhizome; leaves floating or ascending, orbicular, the basal sinus narrow, to 30 (–60) cm in diameter, green above, usually purple beneath, usually pubescent beneath, on flat, glabrous or puberulent, purplish petioles; flowers showy, fragrant, 5–20 cm across, white, rarely pink, long-pedunculate; sepals 4, free, green or sometimes purple on the back, ovate, subacute to obtuse at the apex, to 10 cm long, up to one-third as broad; petals numerous, free, usually white, subacute at the apex, to 3 cm long; seeds ellipsoid to oblongoid, about 2 mm long.

108. Nymphaea odorata (Fragrant Water Lily). *a*. Leaf, X.30. *b*. Flower, X.02¼.

COMMON NAME: Fragrant Water Lily.

HABITAT: Lakes and ponds.

RANGE: Newfoundland to Manitoba, south to Texas and Florida.

ILLINOIS DISTRIBUTION: Rare in Illinois; recorded only from Cook, Franklin, Johnson, McHenry, Perry, and Union counties.

There is some question among botanists whether this species is specifically distinct from *N. tuberosa*. After examining much material of the two, I have decided to recognize both as separate species.

All Illinois specimens examined have white petals, although a pink-flowered form may be expected to occur in Illinois.

Some botanists recognize var. *gigantea* Tricker, a robust taxon with slightly larger flowers and broader leaves which tend to turn up around the edges. At this time, I prefer not to recognize this variety.

The fragrant water lily flowers from June to September.

2. **Nymphaea tuberosa** Paine, Cat. Pl. Oneida Co. N.Y. 184. 1865. *Fig. 109*.

Castalia tuberosa (Paine) Greene, Bull. Torrey Club 15:84. 1888.

Aquatic perennial from a stout, forking rhizome; leaves floating, orbicular, the basal sinus narrow, to 40 cm across, green above and below, glabrous or slightly pubescent below, on stout petioles green and with brown stripes above; flowers showy, not fragrant, to 25 cm across, white, long-pedunculate; sepals 4, free, green, narrowly ovate, subacute to obtuse at the apex, to 10 cm long, up to one-third as broad; petals numerous, free, white, rounded at the apex, to 3 cm long; seeds globose-ovoid, 3–4 mm long.

COMMON NAME: White Water Lily.

HABITAT: Ponds and streams.

RANGE: Quebec to Ontario, south to Nebraska, Arkansas, and Maryland.

ILLINOIS DISTRIBUTION: Occasional in the northern two-thirds of Illinois.

The white water lily is more common in Illinois than *N. odorata*, where it occurs in shallow water.

109. Nymphaea tuberosa (White Water Lily).   *a*. Leaf, X⅓.   *b*. Flower, X¼.

The tuberous rootstocks of this species are an important source of food for beavers, muskrats, and other wildlife.

The white water lily flowers from June to August.

NELUMBONACEAE–LOTUS FAMILY

Only the following genus comprises this family.

1. Nelumbo Adans.–Sacred Bean

Aquatic perennial from thickened rootstocks; leaves large, alternate along the rootstocks; flowers solitary, showy, bisexual, actinomorphic; sepals 4–5, free, green; petals numerous, free, inserted on the calyx; stamens numerous, free, inserted on the calyx; pistils numerous, free, borne in pits in the large receptacle, with 1–2 pendulous ovules; fruits nutlike, sunk in pits in the woody receptacle.

There are three species of *Nelumbo*. In addition to the American species, there is one Jamaican species and one Asian species. The Asian species, *N. nucifera* Gaertn., known as the sacred lotus, is a major source of food in southern China.

Only the following species occurs in Illinois.

1. **Nelumbo lutea** (Willd.) Pers. Syn. 1:92. 1805. *Fig. 110.*

Nymphaea pentapetala Walt. Fl. Carol. 155. 1788. Based on a monstrosity.

Nelumbium luteum Willd. Sp. Pl. 2:1259. 1799.

Nelumbo pentapetala (Walt.) Fern. Rhodora 36:23. 1934.

Aquatic perennial with a tuberous rootstock; leaves floating or raised out of the water, orbicular, green above and below, glabrous above, more or less pubescent beneath, to 60 cm across, peltate, on long glabrous petioles; flowers pale yellow, to 25 cm across, on long, stout peduncles; sepals 4–5, green; petals obovate, concave, obtuse to subacute at the apex; stamens numerous, the anthers tipped by a hooked appendage, the filaments petaloid; fruits more or less hemispheric, woody, up to 20 cm across, with numerous seeds sunk in the receptacle, the seeds globose, to 1.5 cm in diameter.

110. Nelumbo lutea (American Lotus).   *a*. Leaf, X.02¼.   *b*. Flower, X.30.   *c*. Stamen, X2¼.   *d*. Fruit, X.02¼.   *e*. Seed, X.30.

COMMON NAME: American Lotus.

HABITAT: Lakes and ponds.

RANGE: Massachusetts to southern Ontario and Minnesota, south to Texas and Florida.

ILLINOIS DISTRIBUTION: Occasional throughout the state, but becoming very infrequent in the northern part of Illinois.

This handsome aquatic may sometimes form dense colonies of several acres.

The tuberlike rootstocks of this species are rich in starch and can be used as a source of food, as they almost certainly were by the American Indians. The seeds may be roasted and eaten. The rootstocks and seeds are also a source of food for wildlife.

The flowers are borne in July and August.

CABOMBACEAE–CABOMBA FAMILY

Aquatic perennials with rhizomes; stems covered with mucilage; leaves floating or immersed, alternate, opposite, or whorled, sometimes dissected into filiform segments, some or all of them peltate; flowers solitary, axillary, bisexual, actinomorphic; sepals 3 (–4), free, green; petals 3 (–4), free, white, yellow, or purple; stamens 3–18, free; pistils 2–18, free, each with 2–3 ovules; fruit a cluster of 2 or more indehiscent, 1- to 3-seeded follicles.

The family Cabombaceae is closely related to the Nymphaeaceae and Nelumbonaceae, differing from both by its definite number of perianth parts and its follicular fruits. Several botanists combine all three families into one.

KEY TO THE GENERA OF Cambombaceae IN ILLINOIS

1. Leaves uniform; stamens 12–18; carpels 4–18 _____ 1. *Brasenia*
1. Leaves dimorphic; stamens 3–6; carpels 2–3 _____ 2. *Cabomba*

1. *Brasenia Schreb–Watershield*

Aquatic perennial; leaves alternate, centrally peltate, entire; flowers axillary; sepals 3 (–4), free, green; petals 3 (–4), free, purple; stamens 12–18, free; pistils 4–18, free, each with 2–3 pendulous ovules; fruit a cluster of 1- to 2-seeded, coriaceous, indehiscent ripened carpels.

Only the following species comprises the genus.

1. **Brasenia schreberi** Gmel. Syst. Veg. 1:853. 1796. *Fig. 111.*

Brasenia peltata Pursh, Fl. Am. Sept. 389. 1814.

Aquatic perennial from a slender rhizome; leaves floating, oval to nearly orbicular, obtuse at either end, to 10 cm long, up to two-thirds as broad, entire, green and glabrous on both surfaces but covered by a viscid jelly; flowers to 1.5 cm across, on long, glabrous, axillary peduncles; sepals 3 (–4), linear, green; petals 3 (–4), linear, purple; stamens 12–18, free, with long, filiform filaments; fruit a cluster of 4–18 oblongoid follicles to 1 cm long, tipped by the persistent style, 1- to 2-seeded.

COMMON NAME: Watershield.

HABITAT: Ponds, lakes, and quiet streams.

RANGE: Nova Scotia to British Columbia, south to Oregon, Idaho, Texas, and Florida; Alaska; Central America; West Indies; Asia; Australia; Africa.

ILLINOIS DISTRIBUTION: Rare but scattered in the state.

The watershield is widely distributed in much of the world. Although related to *Cabomba*, it is easily distinguished by its uniform leaves, 12–18 stamens, 4–18 pistils, and purple flowers.

Pursh's *Brasenia peltata* is the same species as *B. schreberi*. The flowers open from June to September.

2. *Cabomba Aubl.*–Carolina Watershield

Aquatic perennial; leaves dimorphic, the submersed ones opposite or whorled, dissected into filiform segments, not peltate, the floating ones becoming alternate, entire, centrally peltate; flowers axillary; sepals 3, free, green; petals 3, free, white or yellow; stamens 3–6, free; pistils 2–3, free, each with 3 pendulous ovules; fruit a cluster of 2–3 three-seeded, coriaceous, indehiscent ripened carpels.

Four species native of the tropical and temperate New World comprise the genus.

Only the following species occurs in Illinois.

1. **Cabomba caroliniana** Gray, Ann. Lyc. N. Y. 4:47. 1837. *Fig. 112.*

Aquatic perennial from a slender rhizome; submersed leaves opposite or whorled, not peltate, palmately dissected into many fili-

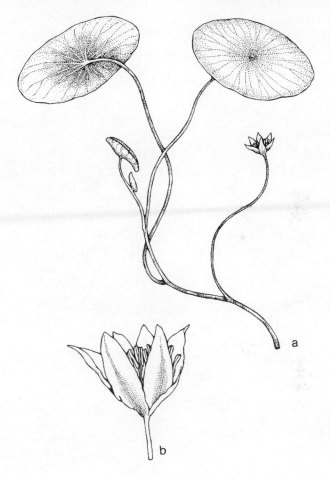

111. Brasenia schreberi (Watershield). *a*. Habit, X½. *b*. Flower, X2½.

form segments, the floating leaves becoming alternate, linear-oblong, entire, to 2 cm long; flowers to 1.8 cm broad, on long, glabrous, axillary peduncles; sepals 3, linear, green; petals 3, obovate, white with usually yellow bases; stamens 3–6, free; fruit a cluster of 2–3 flask-shaped, pubescent follicles to 1 cm long, with 3 seeds.

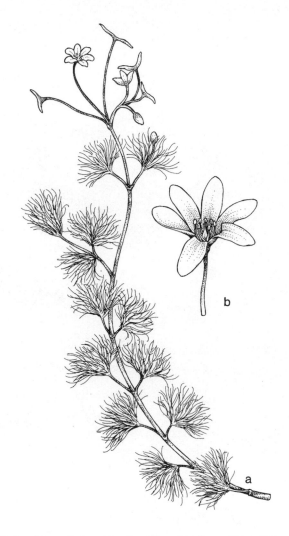

112. Cabomba caroliniana (Carolina Watershield).   *a*. Habit, X½.   *b*. Flower, X2.

COMMON NAME: Carolina Watershield.

HABITAT: Ponds.

RANGE: North Carolina to southern Missouri, south to Texas and Florida; apparently adventive in the northeastern United States.

ILLINOIS DISTRIBUTION: Rare in the southern one-third of the state; also Macoupin, Sangamon, and Vermilion counties.

The Carolina watershield is readily distinguished by its dimorphic leaves, 3 sepals, 3 petals, 3–6 stamens, and cluster of 2–3 follicles.

The flowers appear from May to September.

CERATOPHYLLACEAE–HORNWORT FAMILY

Only the following genus comprises the family.

1. *Ceratophyllum* L.–*Hornwort*

Herbaceous aquatics with much-branched stems; leaves whorled, several-forked; flowers monecious, sessile, solitary in the axils, each subtended by a several-forked involucre; calyx absent; corolla absent; stamens 10–numerous, free; pistil one, the ovary superior, 1-locular, with 1 pendulous ovule; fruit an achene beaked by the persistent style.

The Ceratophyllaceae are closely allied to the Nymphaeaceae and the Nelumbonaceae. They are all aquatics, they all lack vessels, and they have numerous stamens.

The whorled leaves and sessile, axillary flowers distinguish this genus.

There are three species in the genus, one or more of which are found in most of the fresh waters of the world.

The genus has been revised by Fassett (1953).

KEY TO THE SPECIES OF Ceratophyllum IN ILLINOIS

1. Achenes with 2 basal spines; ultimate leaf segments toothed on the margins _____ 1. *C. demersum*
1. Achenes with several spines both lateral and basal; ultimate leaf segments not toothed on the margins _____ 2. *C. echinatum*

1. Ceratophyllum demersum L. Sp. Pl. 992. 1753. *Fig. 113a–c.*

Herbaceous aquatic; stems much branched, glabrous, varying in

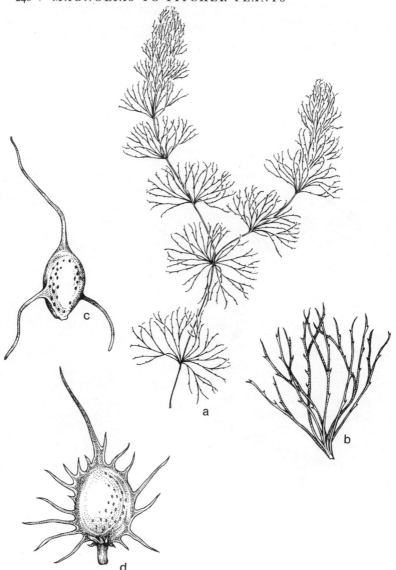

113. Ceratophyllum demersum (Coontail). *a*. Habit, X½. *b*. Branch, X2. *c*. Fruit, X6. *Ceratophyllum echinatum* (Spiny Coontail). *d*. Fruit, X6.

length in accordance with the depth of the water; leaves whorled, sessile, to 3 cm long, glabrous, divided usually into 3-forked, filiform segments, the segments flattened and serrated; flower solitary in the axils, unisexual, subtended by an 8- to 12-fid involucre; stamens 10–20, free, the anthers sessile or nearly so; fruit ellipsoid, compressed, glabrous, wingless, the body 4–5 mm long, the 2 basal spurs 2–5 mm long, the beak 4–6 mm long.

COMMON NAME: Coontail; Hornwort.

HABITAT: Quiet waters.

RANGE: Quebec to British Columbia, south to California, Texas, and Florida; Mexico; West Indies; Europe; Asia.

ILLINOIS DISTRIBUTION: Occasional to common throughout Illinois.

Coontail is often very abundant in quiet waters in most parts of the state. It is a variable and poorly understood species. The length and texture of the stems, the degree of serration of the ultimate segments of the leaves, and characters of the fruit are all variable.

The flowers appear from July to September.

2. Ceratophyllum echinatum Gray, Man. ed. 1, 401. 1848. *Fig. 113d.*

Herbaceous aquatic; stems much branched, glabrous, varying in length in accordance with the depth of the water; leaves whorled, sessile, to 3 cm long, glabrous, divided usually into 3-forked, filiform segments, the segments flattened and not serrated; flower solitary in the axils, unisexual, subtended by an 8- to 12-fid involucre; stamens 10–20, free, the anthers sessile or nearly so; fruit ellipsoid, compressed, minutely tuberculate, narrowly winged, the body 5–7 mm long, the several lateral and basal spines and spurs 2–5 mm long, the beak 5–10 mm long.

COMMON NAME: Spiny Coontail; Spiny Hornwort.

HABITAT: Quiet waters.

RANGE: New Brunswick to southern Quebec and Minnesota, south to Texas and Florida; Mexico.

ILLINOIS DISTRIBUTION: Not common, but scattered throughout the state.

The spiny coontail lives in fresh water, often growing with *C. demersum*. Although it is less common than *C. demersum*, *C. echinatum* probably is more frequent in Illinois than present records indicate, since it is often overlooked by most collectors. The first Illinois collection was made by S. B. Mead in 1848 from Hancock County.

The flowers occur from July to September.

Order Sarraceniales

The only family attributed to this order is the Sarraceniaceae.

SARRACENIACEAE–PITCHER PLANT FAMILY

Perennial herbs; leaves basal, tubular or pitcher-shaped; flower solitary; sepals 4–5, free, persistent; petals 5, free, or absent; stamens numerous, free; pistil one, the ovary superior, 3- to 5-locular, with numerous ovules, the style peltate, usually lobed; fruit a loculicidal capsule, with numerous small seeds.

The Sarraceniaceae are composed of three genera of marsh herbs, all native to the New World.

Only the following genus is represented in the Illinois flora.

1. Sarracenia L.–Pitcher Plant

Perennial herbs; leaves basal, pitcher-shaped or tubular; flower solitary, nodding, subtended by 3 or 4 bracts; sepals 5, free; petals 5, free; stamens numerous, free; pistil one, the ovary superior, 5-locular, with numerous ovules, the style peltate, umbrellalike, with 5 rays, each terminated by hooked stigmas; fruit a 5-locular capsule with many seeds.

Only the following species occurs in Illinois.

1. Sarracenia purpurea L. Sp. Pl. 510. 1753. *Fig. 114.*

Perennial herbs; leaves basal, pitcher-shaped, with a wing on one side and an arching hood at the tip, curved, to 15 cm long, glabrous on the outside, densely clothed on the inside with stiff, reflexed bristles, the tube inflated and hollow, usually green with purple veins; flower solitary on a naked scape, up to 5 cm across; sepals 5, usually purplish, persistent; petals obovate, purple, arched over the greenish-yellow style; style peltate, umbrellalike, 5-lobed; capsule granular, rugose, with many small seeds.

114. Sarracenia purpurea (Pitcher-plant). *a*. Habit, X½.

COMMON NAME: Pitcher Plant.

HABITAT: Bogs.

RANGE: Labrador to Alberta and Minnesota, south to northern Illinois, Maryland, and Delaware.

ILLINOIS DISTRIBUTION: Rare; recorded only from Cook, Lake, and McHenry counties.

The pitcher plant is perhaps the most unique plant in the Illinois flora. The tubular pitcher is hollow and filled with water into which unwary insects fall and drown. Reflexed bristles on the inner surface of the tube prevent captured insects from crawling out.

The strange flowers are produced in May and June.

SPECIES EXCLUDED

Anemone decapetala Ard. Brendel (1887) used this binomial, which applies to a European species, for *A. caroliniana*.

Anemone dichotoma L. This European species does not occur in Illinois, although Brendel (1887) used this binomial for *A. canadensis*.

Anemone nemorosa L. This European species does not occur in Illinois, although Illinois botanists in the nineteenth century used this binomial for *A. quinquefolia*.

Argemone intermedia Sweet. Jones et al. (1955) record this species from three Illinois counties, but I believe the specimens on which they are based are referable to *A. albiflora*.

Delphinium exaltatum Ait. Although the tall larkspur was reported by Brendel in 1859 from Illinois, this is a more eastern species which does not occur in this state.

Nuphar rubrodiscum Morong. This species, reported from Illinois by Pepoon in 1927 as *Nymphaea rubrodiscus* (Morong) Greene, could not be substantiated in this study.

Nuphar sagittifolium (Walt.) Pursh. This taxon, native along the Atlantic coast from Virginia to Florida, was reported as *Nymphaea sagittifolia* Walt. by Jacob Schneck from the Wabash Valley in 1876. I have not been able to verify this record. Although Schneck's identifications are mostly reliable, the fact that *N. sagittifolium* would be far out of range in Illinois makes the report suspect.

Thalictrum cornuti L. Although Mead (1946) and several other botanists during the nineteenth century listed this species from Illinois, the reports were based on misidentifications of *Thalictrum dasycarpum* var. *hypoglaucum*. *Thalictrum cornuti* does not occur in Illinois.

Thalictrum polygamum Muhl. This is an eastern United States species not known from Illinois, although attributed to this state by early workers who were looking at *T. dasycarpum* var. *hypoglaucum*.

Thalictrum purpurascens L. Most Illinois botanists used this binomial for *T. dasycarpum* during the nineteenth century, but this is a different species not in the northeastern United States.

Summary of the Taxa Treated in this Volume

Families	Genera	Species	Lesser Taxa
Magnoliaceae	2	2	
Annonaceae	1	1	
Aristolochiaceae	2	3	2
Calycanthaceae	1	1	
Lauraceae	2	2	2
Saururaceae	1	1	
Menisperma-ceae	3	3	
Ranunculaceae	18	61	8
Berberidaceae	4	6	
Papaveraceae	11	21	
Nymphaeaceae	2	3	1
Nelumbonaceae	1	1	
Cabombaceae	2	2	
Ceratophylla-ceae	1	2	
Sarraceniaceae	1	1	
Totals	52	110	13

ADDENDUM 2017
GLOSSARY
LITERATURE CITED
INDEX OF PLANT NAMES

ADDENDUM 2017

Since the publication of *Flowering Plants: Magnolias to Pitcher Plants* in 1981, several plants in the families included in the book are new to Illinois and have, therefore, been added here. Most of them are escapes from gardens and do not affect the native flora of Illinois. In addition, several name changes have been made, and they are included in this addendum.

MAGNOLIACEAE

Magnolia stellata Maxim. Star magnolia. This is a new record for Illinois. Escape from cultivation. DuPage County, Warrenville Grove Forest Preserve, near old homesite, *Kobal 12–06.*

Magnolia X loebneri Kache. Loebner's magnolia. This is a new record for Illinois. Escape from cultivation. DuPage County, Herrick Lake Forest Preserve, old field, *Lampa & Kobal 94–38.*

Magnolia X soulangeana Soul.-Bod. Saucer magnolia. This is a new record for Illinois. Escape from cultivation. DuPage County, Warrenville Grove Forest Preserve, *Kobal 12–14.*

ARISTOLOCHIACEAE

Asarum canadense L. var. *ambiguum* (Bickn.) Farw. This is a new variety for Illinois. It differs from the other varieties of *A. canadense* by its slightly reflexed calyx lobes 12–20 mm long. Rich woods; northern half of Illinois; native.

Asarum europaeum L. Asarabacca. This is a new record for Illinois. Escape from cultivation. DuPage County, Lyman Woods Nature Center, *Kobal 12–08.*

Aristolochia serpentaria L. var. *serpentaria*. This taxon is now known as *Endodeca serpentaria* (L.) Raf.

Aristolochia serpentaria L. var. *hastata* (Nutt.) Duchartre. I am recognizing this taxon to be a distinct species known as *Endodeca hastata* (Nutt.) Raf.; native.

Aristolochia tomentosa Sims. I now call this plant *Isotrema tomentosa* (Sims) Huber.

Aristolochia clematitis L. This is a new record for Illinois. DuPage County, Lyman Woods Nature Center, old homesite, *Kobal 12–52*; nonnative.

RANUNCULACEAE

Ranunculus repens L. var. *pleniflorus* Fern. is now called *Ranunculus repens* L. var. *degeneratus* Schur.

Ranunculus cymbalaria Pursh. It is my belief now that this species should be placed in the genus *Halerpestes* Greene, thus becoming *Halerpestes cymbalaria* (Pursh) Greene.

Ceratocephala testiculata (Crantz) Besser is new to Illinois since this book was first published. This nonnative species is most often found in campsites. Some botanists call this plant *Ranunculus testiculatus* Crantz. The wide-spreading achenes form an elongated bur.

Delphinium ajacis L. is now called *Consolida ajacis* (L.) Schur.

Delphinium consolida L. is now called *Consolida regalis* L.

Delphinium pubescens DC. is now known as *Consolida pubescens* (DC.) Soo.

Consolida divaricata (Ledeb.) Schrödinger. This bushy larkspur was found in 1973 at a landfill near a former residence at the Morton Arboretum, DuPage County, *Crowley s.n.*; nonnative.

Consolida orientalis (J. Gay) Schrödinger. A single collection of this escape from cultivation was made in 1954 from a garbage dump in Chicago, Cook County, *Eiseman 61.*

Thalictrum pubescens Pursh is new for Illinois. It was found in a wet meadow in the Shawnee National Forest in Hardin County. It is primarily a species of the Appalachian Mountains region; native.

Hepatica nobilis Schreb. var. *obtusa* (Pursh) Steyerm. I am recognizing this as a distinct species called *Hepatica americana* (DC.) Ker.

Hepatica nobilis Schreb. var. *acuta* (Pursh) Steyerm. I am considering this to be a distinct species known as *Hepatica acutiloba* DC.

Isopyrum biternatum (Raf.) Torr. & Gray. The correct name for this species is *Enemion biternatum* Raf.

Anemonella thalictroides (L.) Spach. I am maintaining this binomial for the rue anemone. Other botanists call this plant *Thalictrum thalictroides* (L.) Eaves & Boivin.

Anemone patens L. var. *multifida* Pritz. I now believe that *Anemone patens* should be segregated into the genus *Pulsatilla* Mill. on the basis of its plumose styles and the presence of staminodia. Our plants are *Pulsatilla patens* (L.) Mill. var. *multifida* (Pritz) S. H. Li & Y. H. Luang.

Anemone blanda Schott & Kotschy, known as sapphire anemone, is a garden escape in Illinois. It has been found at the edge of a woodland in 2012 at Greene Valley Forest Preserve, DuPage County, *Kobal 12–21.*

Anemone quinquefolia L. var. *interior* Fern. I believe this taxon with stems that have spreading villous hairs should be recognized as a distinct variety; native.

Clematis dioscoreifolia Levl. & Vaniot should be known as *Clematis terniflora* DC.

Clematis glauca Willd. This yellow-flowered ornamental was found in 2011 at the Morton Arboretum, DuPage County, *Beck & Means 1326IVII*. It was climbing over a rosebush in the China Collection.

Clematis integrifolia L. was collected from open grassy areas in 2007 at the Morton Arboretum, DuPage County, *Sturner JS343*. It is native to Eurasia.

Clematis occidentalis (Hornem.) DC. This species, new to Illinois, was discovered by John Schwegman on algific slopes in 1991 in Jo Daviess County.

Clematis viticella L. This Eurasian species, known as the Italian leather flower, was found in a detention area in 2013 at the Morton Arboretum, DuPage County, *Potenberg & Dorrell KP-6*.

BERBERIDACEAE

Epimedium pinnatum Fisch. should be added to the Illinois flora on the basis of a collection found in a degraded woodland near Wheaton, DuPage County, *Kobal 12–20*; nonnative.

PAPAVERACEAE

Argemone polyanthemos (Fedde) G. Ownbey. The white prickly poppy has been found in Morgan County. It may represent a range extension to the east.

Papaver orientale L. This garden escape is known from a collection of a presumably spontaneous plant in the Chicago area; nonnative.

Papaver pseudo-orientale Medvedev. There is apparently a spontaneous garden escape of this ornamental from the Chicago area; nonnative.

FUMARIACEAE

I am now following others in placing the genera *Adlumia*, *Corydalis*, *Dicentra*, and *Fumaria* in the Fumariaceae rather than the Papaveraceae.

Corydalis sempervirens (L.) Pers. I now believe it should be moved to the genus *Capnoides* Mill. where it is known as *Capnoides sempervirens* (L.) Borkh.

Corydalis campestris (Britt.) Buchholz & Palmer. I am somewhat reluctantly calling this taxon *Corydalis micrantha* (Engelm.) Gray ssp. *australis* (Chapm.) G. Ownbey.

NYMPHAEACEAE

Nuphar luteum (L.) Sibth. & Small ssp. *variegatum* (Engelm. ex Clinton) Beal. I now recognize this plant as a distinct species known as *Nuphar variegatum* Engelm.; native.

Nuphar luteum (L.) Sibth. & Small ssp. *macrophyllum* (Small) Beal. I now recognize this plant as a distinct species known as *Nuphar advena* (Ait.) Ait.f.; native.

NELUMBONACEAE

Nelumbo nucifera Gaertn. The sacred lotus has been found in Lake Vermilion, Vermilion County. It is new for Illinois; nonnative.

GLOSSARY

Achene. A type of one-seeded, dry, indehiscent fruit with the seed coat not attached to the mature ovary wall.

Actinomorphic. Having radial symmetry; regular, in reference to a flower.

Acuminate. Gradually tapering to a point.

Acute. Tapering to a short point.

Adnate. Fusion of dissimilar parts.

Anther. The terminal part of a stamen which bears pollen.

Anthesis. Flowering time.

Apiculate. Abruptly short-pointed at the tip.

Appressed. Lying flat against the surface.

Aril. An appendage of the seed.

Arillate. Having an aril.

Attenuate. Gradually becoming narrowed.

Bibracteate. With two bracts.

Bifid. Two-cleft.

Bilobed. Divided into two lobes.

Bisexual. Referring to a flower which contains both stamens and pistils.

Biternate. Divided into three parts two times.

Bract. An accessory structure at the base of some flowers, usually appearing leaflike.

Bracteole. A secondary bract.

Bractlet. A small bract.

Bulbous. Swollen.

Caducous. Falling away very early.

Calyx. The outermost group of structures of a flower, composed of sepals.

Campanulate. Bell-shaped.

Canescent. Grayish-hairy.

Capsule. A dry, dehiscent fruit composed of more than one carpel.

Carpel. A simple pistil, or one member of a compound pistil.

Carpellate. Possessing carpels.

Caudex. The woody base of a perennial plant.

Cauline. Belonging to a stem.

Claw. A narrow, basal stalk, particularly of a petal.

Cleistogamous. Hidden.

Compressed. Flattened.

Concave. Curved on the inner surface; opposite to convex.

Connective. That portion of the stamen between the two anther halves.

Connivent. Coming in contact; converging.

Convex. Rounded on the outer surface; opposite to concave.

Cordate. Heart-shaped.

Coriaceous. Leathery.

Corolla. The group of structures of a flower just within the calyx, composed of petals.

Corymb. A type of inflorescence where the pedicellate flowers are arranged along an elongated axis but with the flowers all attaining about the same level.

Corymbose. Bearing a corymb.
Crenate. Round-toothed.
Crested. Bearing a ridge.
Crisped. Curled.
Cuneate. Wedge-shaped or tapering at the base.
Cyme. A type of broad and flattened inflorescence in which the central flowers bloom first.
Cymose. Bearing a cyme.

Deciduous. Falling away.
Decompound. More than once compound or divided.
Dehiscent. Splitting at maturity.
Deltoid. Triangular.
Dentate. With sharp teeth, the tips of which project outward.
Denticulate. With small, sharp teeth, the tips of which project outward.
Depressed. Flattened from above.
Dimorphic. Having two forms.
Dioecious. With staminate flowers on one plant, pistillate on another.
Dissected. Divided into narrow segments.
Drupe. A type of fruit in which the seed is surrounded by a hard, dry covering which, in turn, is surrounded by fleshy material.

Eglandular. Without glands.
Ellipsoid. Referring to a solid object which is broadest at the middle, gradually tapering equally to both ends.
Elliptic. Broadest at the middle, gradually tapering equally to both ends.
Entire. Bearing no teeth along the margins.

Excentric. Off-center.
Exserted. Projecting beyond.

Fibrous. Referring to roots borne in tufts.
Filament. That part of the stamen supporting the anther.
Filiform. Threadlike.
Follicle. A type of dry, dehiscent fruit which splits along one side at maturity.
Furrow. A groove.
Fusiform. Spindle-shaped; tapering at both ends.

Glabrate. Becoming smooth.
Glabrous. Without pubescence or hairs.
Gland. A structure that secretes.
Glandular. With glands.
Glaucous. With a whitish covering that can be rubbed off.
Globose. Round; globular.
Granular. Grainy in texture.

Hastate. Spear-shaped; said of a leaf which is triangular with spreading basal lobes.
Hirsute. With stiff hairs.
Hirsutulous. With minute stiff hairs.
Hirtellous. Finely hirsute.
Hispid. With rigid hairs.
Hispidulous. With minute rigid hairs.

Imbricate. Overlapping.
Indehiscent. Not splitting open at maturity.
Inferior. Referring to the position of the ovary when it is surrounded by the adnate portion of the floral tube or is embedded in the receptacle.

Inflorescence. A cluster of flowers.

Involucral. Pertaining to the involucre.

Involucre. A circle of bracts which subtends a flower cluster.

Irregular. In reference to a flower having no symmetry at all.

Keel. A ridge.

Lanceolate. Lance-shaped; broadest near base, gradually tapering to the narrower apex.

Lanceoloid. Referring to a solid object which is broadest near base, gradually tapering to the narrower apex.

Linear. Narrow and approximately the same width at either end and the middle.

Locular. Referring to the cells or cavities of a compound ovary.

Locule. A cell or cavity of a compound ovary.

Loculicidal. Said of a capsule which splits down the dorsal suture of each cell.

Lustrous. Shiny.

Membranaceous. Thin and more or less transparent.

Monoecious. Bearing both sexes in separate flowers on the same plant.

Mucronulate. Possessing a very short, abrupt tip.

Nectariferous. Having nectaries.

Nectary. Any structure where nectar is produced.

Obcuneate. Reverse wedge-shaped.

Oblanceolate. Reverse lance-shaped; broadest at apex, gradually tapering to narrower base.

Oblong. Broadest at the middle, and tapering to both ends, but broader than elliptic.

Oblongoid. Referring to a solid object which, in side view, is nearly the same width throughout, but broader than linear.

Obovate. Broadly rounded at the apex, becoming narrowed below.

Obovoid. Referring to a solid object which is broadly rounded at the apex, becoming narrowed below.

Obtuse. Rounded at the apex.

Orbicular. Round.

Oval. Broadly elliptic.

Ovate. Broadly rounded at base, becoming narrowed above; broader than lanceolate.

Ovoid. Referring to a solid object which is broadly rounded at the base, becoming narrowed above.

Ovule. The egg-producing structure found within the ovary.

Palmate. Divided radiately, like the fingers of a hand.

Panicle. A type of inflorescence composed of several racemes.

Paniculate. Bearing a panicle.

Papillate. Bearing pimplelike processes.

Pedicel. The stalk of a flower of an inflorescence.

Pedicellate. Having pedicels.

Peduncle. The stalk of an inflorescence.

Pedunculate. Having a peduncle.

Peltate. Attached away from the margin, in reference to a leaf.

Pendent. Suspended; overhanging.

Pendulous. Hanging.

Perfect. Bearing both stamens and pistils in the same flower.

Perianth. Those parts of a flower including both the calyx and corolla.

Petal. One segment of the corolla.

Petaloid. Resembling a petal in texture and appearance.

Petiolate. Having a petiole.

Petiole. The stalk of a leaf.

Petiolulate. Having petiolules.

Petiolule. The stalk of a leaflet.

Pilose. Bearing soft hairs.

Pilosulous. Bearing soft, short hairs.

Pinnate. Divided once into distinct segments.

Pinnatifid. Said of a simple leaf which is cleft or lobed only part way to its axis.

Pistil. The ovule-producing organ of a flower normally composed of an ovary, a style, and a stigma.

Pistillate. Bearing functional pistils but not stamens.

Plumose. Having fine, elongated hairs.

Polygamous. With both perfect and unisexual flowers.

Puberulent. With minute hairs.

Pubescent. Bearing some kind of hairs.

Pyriform. Pear-shaped.

Quadrate. Nearly square.

Raceme. A type of inflorescence where pedicellate flowers are arranged along an elongated axis.

Racemose. Bearing a raceme.

Rachis. The axis of the inflorescence.

Receptacle. That part of the flower to which the perianth, stamens, and pistils are usually attached.

Reflexed. Turned downward.

Regular. Having radial symmetry (actinomorphic) or bilateral symmetry (zygomorphic).

Reinform. Kidney-shaped.

Reticulate. Resembling a network.

Retrorse. Pointing downward.

Retuse. Shallowly notched at a rounded apex.

Revolute. Rolled under from the margin.

Rhizomatous. Bearing rhizomes.

Rhizome. An underground horizontal stem, bearing nodes, buds, and roots.

Rugose. Wrinkled.

Rugulose. Minutely wrinkled.

Sagittate. Shaped like an arrowhead.

Samara. A winged seed.

Scape. A leafless stalk bearing a flower or inflorescence.

Scapose. Possessing a scape.

Sepal. One segment of the calyx.

Serrate. With teeth, the tips of which project forward.

Serrulate. With very small teeth, the tips of which project forward.

Sessile. Without a stalk.

Sinuate. Wavy along the margins.

Spatulate. Oblong, but with the basal end elongated.

Spherical. Round.

Spike. A type of inflorescence where sessile flowers are arranged along an elongated axis.

Spur. A saclike extension of the flower.

Stamen. The pollen-producing organ of a flower composed of a filament and an anther.

Staminate. Bearing functional stamens but not pistils.

Staminodium (pl., **staminodia**). A sterile stamen.

Stigma. The terminal, pollen-receiving part of the pistil.

Stipitate. Bearing a stalk.

Stipule. A leaflike or scaly structure found at the point of attachment of a leaf to the stem.

Stolon. A slender horizontal stem on the surface of the ground.

Stoloniferous. Bearing stolons.

Strigose. With appressed, straight hairs.

Style. That part of the pistil between the ovary and the stigma.

Subcoriaceous. Somewhat leathery.

Subinferior. Referring to the position of the ovary when the ovary is partially embedded in the receptacle.

Subulate. Drawn to an abrupt short point.

Succulent. Fleshy; juicy.

Superior. Referring to the position of the ovary when the free floral parts arise below the ovary.

Tendril. A spiraling, coiling structure which enables a climbing plant to attach itself to a supporting body.

Terete. Round in cross section.

Ternate. Divided three times.

Tomentose. Pubescent with matted wool.

Torulose. With small constrictions.

Trifoliolate. Having three leaflets.

Truncate. Abruptly cut across.

Tuberculate. With small warts.

Tubular. Shaped like a tube.

Turbinate. Top-shaped.

Umbel. A type of inflorescence in which the flower stalks arise from the same level.

Undulate. Wavy.

Unisexual. Bearing either stamens or pistils in one flower.

Valvate. Opening by valves.

Valve. That part of a capsule or anther which splits open.

Villous. With long, soft, slender, unmatted hairs.

Whorl. An arrangement of three or more structures at a point on the stem.

Zygomorphic. Bilaterally symmetrical.

LITERATURE CITED

Bailey, W. M. 1949. Initial report on the vascular plants of southern Illinois. Transactions of the Illinois Academy of Science 42:47–55.

Beal, E. O. 1956. Taxonomic revision of the genus *Nuphar* Sm. of North America and Europe. Journal of the Elisha Mitchell Society 72:317–46.

Beck, L. C. 1828. Contributions towards the botany of the states of Illinois and Missouri. American Journal of Science & Arts 14:112–21.

Benson, L. 1948. A treatise on the North American Ranunculi. American Midland Naturalist 40:1–261.

Boivin, B. 1957. Reduction du genre *Anemonella* Spach. Bulletin de la Société Royale de Botanique de Belgique 89:319–21.

Brendel, F. 1859. Additions and annotations to Mr. Lapham's catalogue of Illinois plants. Transactions of the Illinois State Agricultural Society 3:583–85.

———. 1887. Flora Peoriana. Privately printed for the author at Peoria, Ill. 89 pp.

Cowles, H. C. 1901. The physiographic ecology of Chicago and vicinity. Botanical Gazette 31:73–108, 145–82.

Cronquist, A. 1968. The evolution and classification of flowering plants. Boston: Houghton Mifflin Co. 396 pp.

Fassett, N. C. 1953. North American *Ceratophyllum*. Comunicaciones Instituto Tropical de Investigaciones Cientificas 2:25–45.

Fernald, M. L. 1938. *Ranunculus abortivus* and its eastern American allies. Rhodora 40:416–20.

———. 1940. Some spermatophytes of eastern North America. Rhodora 42:239–76.

———. 1950. Gray's manual of botany. 8th ed. New York: American Book Co. 1632 pp.

Forbes, S. A. 1870. Botanical notes. American Entomologist and Botanist 2:310.

Gleason, H. A. 1952. The new Britton and Brown illustrated flora of the northeastern United States and adjacent Canada. New York: New York Botanical Garden. Vol. 2. 655 pp.

Higley, W. K., and C. S. Raddin. 1891. Flora of Cook County, Illinois and a part of Lake County, Indiana. Bulletin of the Chicago Academy of Science 2:1–168.

Hutchinson, J. 1973. The families of flowering plants. Oxford: Clarendon Press. 968 pp.

Jones, G. N., G. D. Fuller, G. S. Winterringer, H. E. Ahles, and A.

Flynn. 1955. Vascular plants of Illinois. Urbana: University of Illinois Press, and the Illinois State Museum, Springfield. 593 pp.

Kral, R. 1960. A revision of *Asimina* and *Deeringothamnus* (Annonaceae). Brittonia 237–78.

Lapham, I. A. 1857. Catalogue of the plants of the state of Illinois. Transactions of the Illinois State Agricultural Society 2:492–550.

Mead, S. B. 1846. Catalogue of plants growing spontaneously in the state of Illinois, the principal part near Augusta, Hancock County. Prairie Farmer 6:35–36, 60, 93, 119–22.

Mohlenbrock, R. H. 1975. Guide to the vascular flora of Illinois. Carbondale: Southern Illinois University Press. 494 pp.

———. 1978. Hollies to Loasas. The Illustrated Flora of Illinois. Carbondale and Edwardsville: Southern Illinois University Press. 315 pp.

Munz, P. A. 1946. *Aquilegia*: the cultivated and wild columbines. Gentes Herbarium 7:1–150.

Ownbey, G. B. 1947. Monograph of the North American species of *Corydalis*. Annals of the Missouri Botanical Garden 34:187–260.

———. 1958. Monograph of the genus *Argemone* for North America and the West Indies. Memoirs of the Torrey Botanical Club 21:1–159.

Patterson, H. N. 1876. Catalogue of the phaenogamous and vascular cryptogamous plants of Illinois. Privately printed by the author at Oquawka, Ill. 54 pp.

Pepoon, H. S. 1927. An annotated flora of the Chicago area. Bulletin of the Chicago Academy of Science 8:1–554.

Ramsey, G. W. 1964. A biosystematic study of the genus *Cimicifuga* (Ranunculaceae). Ph.D. dissertation, University of Tennessee. 279 pp.

Ridgway, R. 1872. Notes on the vegetation of the Lower Wabash Valley. American Naturalist 6:658–65.

Schneck, J. 1876. Catalogue of the flora of the Wabash Valley. Annual Report of the Geological Survey of Indiana 7:504–79.

Shinners, L. 1950. Notes. Field and Laboratory 18:42.

Steyermark, J. A. and C. S. Steyermark. 1960. *Hepatica* in North America. Rhodora 62:223–32.

Swink, F. 1974. Plants of the Chicago region. 2d ed. Lisle, Ill.: Morton Arboretum. 474 pp.

Tehon, L. R. 1942. Fieldbook of native Illinois shrubs. Urbana: Illinois Natural History Survey, Manual No. 3. 307 pp.

Thorne, R. F. 1968. Synopsis of a putatively phylogenetic classification of the flowering plants. Aliso 6:57–66.

Vasey, G. 1870. Some plants to name. American Entomologist and Botanist 2:256.

INDEX OF PLANT NAMES